T0136875

Smart Network Inspired Paradigm and Approaches in IoT Applications

Mohamed Elhoseny · Amit Kumar Singh
Editors

Smart Network Inspired Paradigm and Approaches in IoT Applications

Editors
Mohamed Elhoseny
Faculty of Computers and Information
Mansoura University
Mansoura, Egypt

Amit Kumar Singh
Department of Computer Science
and Engineering
National Institute of Technology Patna
Patna, India

ISBN 978-981-13-8616-9   ISBN 978-981-13-8614-5 (eBook)
https://doi.org/10.1007/978-981-13-8614-5

This Springer imprint is published by the registered company Springer Nature Singapore Pte Ltd.
The registered company address is: 152 Beach Road, #21-01/04 Gateway East, Singapore 189721, Singapore

Preface

Smart networks' architectures and technologies play an important and increasing role in practice due to the widespread adoption of mobile devices in many applications. For example, from the industry perspective, the synergy between smart mobile networks and cloud technologies has resulted in new cloud provisioning models for supporting mobile application development and deployment. From an academic perspective, smart networks are a way of augmenting mobile devices and dealing with the inherent limitations related to remote resources located in IoT applications. Techniques materializing this idea include off-loading and cyber foraging. While smart networks can be viewed as a special case of ad hoc and sensors networks, it represents also an evolution of the latter since it includes the ability of augmenting mobile (e.g., laptops, smartphones, tablets, and wearables) and wireless devices (e.g., sensors and actuators) with processing/storage resources in their proximity, in terms of network topology. Indeed, several flavors of this idea, including micro-data centers, cloudlets, and fog computing itself, follow the edge computing model, by which data/computations are processed using computing resources located at the edge of the network—accessible through wireless protocols —and optionally using remote resources in the cloud.

Motivated not only by the increasing number of mobile devices, but also by their ever-growing computing and sensing capabilities, there have been efforts to leverage these devices as a destination for off-loading computations/data in the context of IoT applications. Such a trend has also been referred to as dew computing in the literature. However, current research in the area is still focused on augmenting mobile clients via fixed computing resources (e.g., local servers and computer clusters), so huge unexploited computing and sensing capabilities remain "at the edge." Therefore, many research opportunities to exploit mobile devices in the context of smart networks for IoT applications arise.

In view of the above, this book presents the state-of-the-art techniques and approaches, design, development, and innovative use of smart networks' inspired paradigm and approaches in IoT as well as other demanding applications. Various recent algorithms and novel/improved techniques are discussed in this book.

The authors believe that this book would provide a sound platform for understanding the smart networks' inspired paradigm and approaches/issues in emerging applications, prove as a catalyst for researchers in the field, and shall be equally beneficial for professionals. In addition, this book is also helpful for the senior undergraduate and graduate students, researchers, and industry professionals working in the area as well as other emerging applications demanding smart networks.

We are sincerely thankful to all the authors, editors, and publishers whose works have been cited directly/indirectly in this manuscript.

Special Acknowledgements

The first editor gratefully acknowledges the authorities of *Mansoura University, Egypt*, for their kind support to complete this book.

The second editor gratefully acknowledges the authorities of *National Institute of Technology Patna, India*, for their kind support to come up with this book.

Mansoura, Egypt Dr. Mohamed Elhoseny
Patna, India Dr. Amit Kumar Singh

Contents

About the Editors

Dr. Mohamed Elhoseny is currently an Assistant Professor at the Faculty of Computers and Information, Mansoura University, Egypt, where he is also Director of the Distributed Sensing and Intelligent Systems Lab. Dr. Elhoseny has authored or co-authored over 100 ISI journal articles, conference proceedings, book chapters, and several books published by Springer and Taylor & Francis. His research interests include Network Security, Cryptography, Machine Learning Techniques, and Intelligent Systems. Dr. Elhoseny serves as the Editor-in-Chief of the International Journal of Smart Sensor Technologies and Applications, and as an Associate Editor of several journals such as IEEE Access.

Amit Kumar Singh received the bachelor's degree and M.Tech. in computer science and engineering from Veer Bahadur Singh Purvanchal University, Jaunpur, India (2005) and Jaypee University of Information Technology, Waknaghat, India (2010), prior to completing his Ph.D. degree in computer engineering at the National Institute of Technology, Kurukshetra, India (2015). He worked at the Computer Science and Engineering Department, Jaypee University of Information Technology, from 2008 to 2018. He is currently an Assistant Professor at the Computer Science and Engineering Department, National Institute of Technology at Patna (An Institute of National Importance), Patna, India. Dr. Singh has authored over 80 peer-reviewed journal articles, conference publications, and book chapters, as well as the books Medical Image Watermarking: Techniques and Applications (2017), and Animal Biometrics: Techniques and Applications (2018, Springer International Publishing). He has also edited several books and currently serves on the Editorial Board of two peer-reviewed international journals. His research interests include Data Hiding, Biometrics, and Cryptography.

A Reactive Hybrid Metaheuristic Energy-Efficient Algorithm for Wireless Sensor Networks

N. Shivaraman and S. Mohan

Abstract Expanding network lifespan is the main target during the design of a wireless sensor network. Clustering the sensor nodes is an efficient topology to accomplish this objective. In this work, we offer a reactive hybrid protocol to enhance network lifetime using the hybridization of ant colony optimization (ACO) along with particle swarm optimization (PSO) algorithm. In order to improve the energy efficiency, the anticipated RAP algorithm uses a reactive data transmission strategy which is incorporated into the hybridization of ACO and PSO algorithm. In the beginning, the clusters are organized depending on the residual energy and then the proposed RAP algorithm will be executed to improvise the inter-cluster data aggregation and reduces the data transmission. The experimental outcomes demonstrate the proposed RAP algorithm performs well against other near conventions in different situations.

Keywords WSN · Clustering · Energy efficiency · ACO · PSO

1 Introduction

A wireless sensor network (WSN) is an autonomous wireless network framework comprising of various sensors, which assemble data from their encompassing surroundings as well as broadcast it to an information sink or a base station (BS) [1]. In WSN applications, the primary target is to screen and gather sensor information as well as afterward broadcast information to the BS. Sensors in various areas of the field are able to work together in information accumulation, in addition, to give better precise reports over their neighborhood areas. Mostly, WSNs measures the physical phenomena such as temperature, pressure, acoustic, or area of objects [2], to enhance

N. Shivaraman (✉)
Department of Computer and Information Science, Annamalai University, Chidambaram, India
e-mail: ramansiva28@gmail.com

S. Mohan
Department of Computer Science and Engineering, Annamalai University, Chidambaram, India
e-mail: mohancseau@gmail.com

M. Elhoseny and A. K. Singh (eds.), *Smart Network Inspired Paradigm and Approaches in IoT Applications*, https://doi.org/10.1007/978-981-13-8614-5_1

the accuracy of detailed estimations, as well as information collection decreases the communication congestion in the network, prompting huge energy saving [3–6]. The qualities of minimal effort, low-control, as well as multifunctional sensors have rendered WSNs extremely appealing in different dimensions [7, 8]. These days, with the improvement of cloud innovation [9], WSNs quickly conveyed numerous viable applications, comprising home security, battle-zone reconnaissance, observing the development of undomesticated creatures in the forest, healthcare applications [10, 11], and so forth [12]. Within a sensor network, every node can be a sensor as well as a router. In addition to its computing capability, storage capacity, as well as communication ability, are also the constraints of WSN [13–15]. In addition, in numerous WSN applications, sensor nodes are dispensed into a harsh condition, that makes the substitution of failed nodes moreover troublesome otherwise costly [16–18]. In this way, in numerous situations, a wireless node has to work without battery recharging for an expanded time frame [19]. Thus, energy efficiency is the major issue while planning a WSN by means of the goal of improved network lifespan [7, 20, 21]. Energy utilization might be able to be effectively overseen via modifying the network topology as well as managing the nodes' transmission control in the routing convention [22, 23].

The clustering method is helpful in diminishing the amount of data transmission as well as save energy consumption [24]. Within a clustering design, sensor nodes be formed into clusters, wherever the sensor nodes by means of bring down energy may be able to be utilized to carry out detecting undertakings, and send the detected information to their cluster head (CH) in the nearby distance [25–28]. The clustering model is shown in Fig. 1. A node inside a cluster might be able to be selected as CH to collect the information out of the member of the cluster, through the target of decreasing the transmission distance of the accumulated information transferred to the BS [29, 30].

The clustering method can build network longevity as well as enhance energy proficiency by limiting energy utilization as well as adjusting energy utilization among the nodes [31, 32]. Also, it is equipped for minimizing channel contention

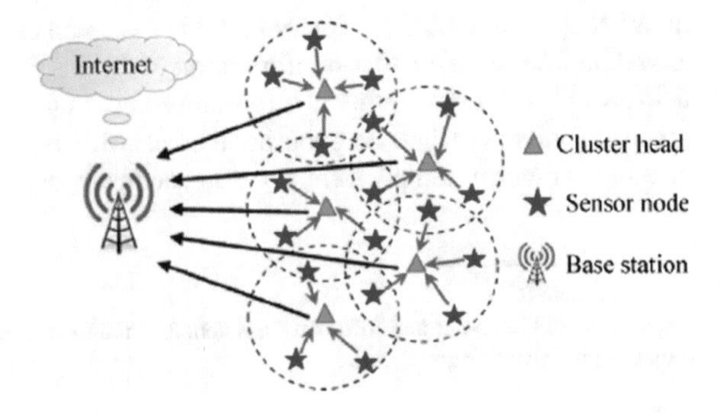

Fig. 1 Clustering architecture

as well as packet collisions, bringing about enhanced network throughput beneath high load [33, 34]. In a standard point of view, the process of optimization can be described to find the best solution of the function from the system within constraints [35–38]. The principal objective of the swarm and evolutionary algorithms (EA) depends on clustering conventions to powerfully cluster sensor nodes in the setup stage so the energy utilization is limited as well as clustering criteria be improved. For N sensor nodes, here are absolutely 2^{N-1} distinctive arrangements, wherever in every arrangement, every node may be able to be whichever chosen as a CH or not. Accordingly, clustering within WSNs is an NP-hard issue. EA in addition to swarm intelligence algorithms contain connected effectively to an assortment of NP-hard issues. Additionally, with regards to clustered routing within WSN, a few EA methods are present in the literature. GCA (genetic clustering algorithm) is an EA for the dynamic development of clusters in WSNs [39]. The goal of GCA is to expand the lifespan of the network by means of comprising least energy scattering. Within GCA, every chromosome is spoken to as a string of length N, wherever every gene relates to a specific node. The estimation of every gene is able to be +1 (showing the node be a CH), 0 (demonstrating the node be a non-CH), or −1 (showing which the node former departed). Jin et al. [40] have proposed another clustering strategy depending on GA endeavor to discover suitable CHs to limit the clustering distances. Within this method, binary encoding is utilized for the chromosome portrayal, in that, every gene relates to one node: 1 implies which comparing node is chosen as a CH, and something else, 0 implies which it is a non-CH node.

To achieve energy efficiency in WSN [41, 42], a new RAP protocol is introduced to enhance the network lifetime using the hybridization of ant colony optimization (ACO) as well as particle swarm optimization (PSO) algorithm. To further enhance the energy effectiveness, the projected RAP algorithm uses a reactive data transmission strategy that is incorporated into the hybridization of ACO and PSO algorithm. In the beginning, the clusters are organized based on the residual energy and then the proposed RAP algorithm will be executed to improvise the inter-cluster data aggregation and reduces the data transmission. In summary, the contributions of the paper are listed below.

- The investigation of the state-of-the-art clustering techniques in WSN takes place.
- From the extensive survey, it is concluded that the efficient data aggregation technique using the hybridization of ACO and PSO algorithms leads to better lifetime.
- A reactive data transmission scheme is also employed to broadcast data alone while the sensed value crosses a threshold limit.
- Simulation results have been analyzed to validate the efficiency of the RAP protocol.

The upcoming sections of the study are formulated as below: The network design is known in Sect. 2, as well as the proposed RAP algorithm, is presented in Sect. 3. The consequences are discussed in Sect. 4 as well as the paper is completed in Sect. 5.

2 Network Model

Some of the assumption made in the proposed work is listed below:

- N sensor nodes are dispensed in the area of M × N.
- Nodes and BS are not mobile.
- Node has its unique ID.
- Symmetric links.

2.1 Energy Model

During the transmission and reception of data, the node has to utilize some energy on the basis of two-channel propagation model known as free space (D2 power loss) in case of single-hop communication and multipath fading channel (D4 power loss) in case of multi-hop communication. So, the energy utilization of the nodes for broadcasting and getting a k-bit packet about a distance d can be represented as

$$E_{TX}(k,d) = \begin{cases} k \times E_{elec} + k \times \varepsilon_{fs} \times d^2 \; if \; d \leq d_0 \\ k \times E_{elec} + k \times \varepsilon_{mp} \times d^4 \; if \; d > d_0 \end{cases} \quad (1)$$

$$E_{RX}(k) = l \times E_{elec} \quad (2)$$

where E_{elec} is the dissipated energy in transmitter otherwise receiver circuit, d_0 is the threshold distance that it is estimated through $\sqrt{\varepsilon_{fs}/\varepsilon_{mp}}$. Depending on the broadcast distance d, free space (ε_{fs}) otherwise multipath fading (ε_{mp}) be employed within the transmitter amplifier.

2.2 Cluster Head (CH) Formation

Thus, this study employs the level flanked clustering process where CHs are elected based on a thresholds function. It defines which node by means of higher energy has a high possibility of becoming CHs. Every node will generate a value randomly and tries to turn out to be CH. When the arbitrary value is lower compared to the threshold T (i), it would turn out to be CH. And, T (i) can be represented as

$$T(i) = \frac{P_{opt}}{1 - p_{opt\left(r.mod\left(\frac{1}{P_{opt}}\right)\right)}} * \frac{E_i(r)}{E_{avg}(r)} \quad (3)$$

where r indicates the present round in WSN, E is the present energy of ith node, whereas E. denotes the average residual energy which can be computed as

$$E_{avg} = \frac{\sum E_i(r)}{N} \quad (4)$$

N indicates the count of nodes within WSN.

3 Proposed RAP Algorithm

The RAP algorithm is employed to locate the shortest route out from CHs to BS. ACO and PSO algorithm [43, 44] has the capability to identify the optimized path among a collection of nodes and BS as the destination. Next, reactive strategy reduces the count of data transmission by allowing the data transmission alone while the sensed value is greater than the threshold value.

3.1 *ACO Based Path Selection*

Here, the least cost based spanning tree (shortest path) is constructed among CHs as well as BS. The steps involved in this algorithm are as follows:

1. Initialization of CHs as ants integrated to BS as the target.
2. Using virtual ant based upon the quantity of pheromone on the CH distances.
3. The beginning of ACO can be the process of collecting trail between nearby clusters, where a number of synthetic ants (CHs) be designed from CHs to BS.
4. At the forefront, ants are selecting the subsequent CH in the random manner by gathering the data out of the length matrix in addition to the successful ants updates the pheromone deposition on the boundaries met by those through an amount (CL), wherever M be the sum of travel time of the ant as well as D a constant price which is altered in continuation by means of the new troubles to the perfect value.
5. The subsequent group of ants follows the leftover pheromone deposition feedback through the previously visited successful ants and quickly follows the shortest route.
6. While ants move from one CH_i to CH_j, the possibility in the selection standard (so known as pheromone) for a simple ant be calculated as follows:

$$P_{ij} = \frac{\left(\tau_{ij}\right)^{\alpha}\left(\eta_{ij}\right)^{\beta}}{\sum_{j\in N}\left(\tau_{ij}\right)^{\alpha}\left(\eta_{ij}\right)^{\beta}} \quad (5)$$

Here, τ_{ij} indicates the quantity of pheromone deposition from the CH_i to CH_j demonstrates the sum of pheromone deposit from CH_i to CH_j. η_{ij} is the trail visibility function which is equal to the reciprocal function of the energy distance between CH_i and CH_j, α and β are the parameters to alter the pheromone quantity of τ_{ij} and η_{ij}, respectively.

7. When a link is available between the CH_i to CH_j; then

   P_{ij} gets updated
   else
   $P_{ij} = 0$
   End

8. The distance d_{ij} between CH_i and CH_j can be calculated as follows:

$$d_{ij} = \sqrt{(S(i).xd - s(j)xd)^2 - (S(i).yd - s(j)yd)^2} \tag{6}$$

   Here, xd and yd are the XY coordinates of the given CH.

9. P values get restructured by every ant that has arrived at the BS.
10. Pheromone evaporation ρ on the edge flanked by CH_i along with CH_j be computed employing the Eq. (7).

$$\tau_{ij} \leftarrow (1 - \rho)\tau_{ij} \tag{7}$$

11. For the CHs which are not selected through artificial ants; the measure of P will decrease in an exponential way.
12. In each round (t) = {1, 2, 3, 4…n}, when each and every ant reaches the BS, the value of τ_{ij} will be equated as
    $\tau_{ij}(t + n) = \rho \cdot \tau_{ij}(t) + \Delta\tau_{ij}$. Here, $\Delta\tau_{ij}$ indicates the pheromone quantity getting settled.
13. If ant k have crossed a few edges between CHs, it would depart P that is indirectly relational to the whole distance end to end of every edges ant k have conceded out off the initial CH to the BS through the use of Eq. by using the following formula:

$$\tau_{ij} \leftarrow \tau_{ij} + \sum_{k=1}^{m} \Delta\tau_{ij}^{k}, \quad \forall (i, j) \in L \tag{8}$$

    Here, $\Delta\tau_{ij}^{k}$ is the quantity of P ant k deposited over the visited limits. It is estimated via the following expression:

$$\Delta\tau_{ij}^{k} = \begin{cases} \frac{1}{C^K} \\ 0 \end{cases} \tag{9}$$

14. At present, the path by means of the best P value is chosen as well as assigned as an initial solution.
15. Finally, PSO algorithm will be executed to reduce the path cost again.

3.2 PSO Algorithm

The initialization of PSO begins with the output of the ACO algorithm as particles. Every particle holds the saved data for every coordinate that is associated to obtain the optimized solution by subsequent presented best particles. The goal function of each particle is validated and saved. The fitness range of the present optimal particle is known as pBest. While each and every created population are taken, after that the best range is selected from the created population as well as the specific best solution is known as gBest. This work utilizes the shortest path cost as goal function. In general, PSO tries to modify the speed of each particle to its pBest. The speed is computed by arbitrary definitions that are actually arbitrarily formed counts for velocity to pBest. Each examined particle of PSO contains the data that is given as follows:

- A data represents a global solution that is named as gBest.
- The rate for velocity would represent the quantity of data to be altered.
- pBest value.

1. Initially, it is considered that every CHs as particle that has two dimensions like particle position as well as velocity.
2. Then the solutions are initiated on the basis of random distribution. And, the count of the random solution depend on the population size.
3. Now, determining the fitness value takes place by a fitness function that is the lowest path distance. The distance flanked by two nodes can be computed as

$$D = \sqrt{(x_1 - x_2)^2 - (y_1 - y_2)^2} \tag{10}$$

   (x_1, y_1) is the position values of node 1 as well as (x_2, y_2) is the location rates of node 2. Once the distance is determined, then it is needed to compute the gBest that is shortest aggregate distance for each arbitrary result.
4. Production of fresh particles out off the early locate of random solutions. Arrangement of fresh particles from the elderly ones is the production of a fresh particle:

   4.1. Estimating new velocity:
   The present velocity in use particle is assumed to be the value at that the particle's location is modified and the fresh velocity can be computed as

$$new_v = \omega * old_v + \omega_1 (lBest_p - cBest_p) + \omega_2 + w_2 (gBest_p - cBest_p) \tag{11}$$

   where ω indicates the inertia weight. ω1 and ω2 are fundamental PSO tuning variables, v indicates the velocity, and p is the location value.

   4.2. Estimating the new position of the particle is as follows:

$$new_p = old_p + new_v \tag{12}$$

Finally, the fresh particle (new_v and new_p) will be obtained.

5. Now, the fitness value for new_p is computed by the use of distance of the path.
6. The fitness value of old particle, as well as new particle, undergoes comparison as well as the top one will be chosen for the subsequent iteration:

$$\begin{aligned}&\textit{If } new_{fv} > old_{fv}\\&old_{fv} = old_{fv}\\&\textit{else}\\&\textit{old particle is forwarded to next iteration}\\&\textit{end}\end{aligned}$$

7. In each iteration, the best solution will be chosen as pBest. The particle with high fitness value in the present iteration is chosen as pBest solution.
8. The pBest solutions every iterations of the particle in that has highest in the midst of all solutions is chosen as gBest solution. In the end, the gBest solution is selected as the present inter-cluster data aggregation route.

3.3 Reactive Data Transmission

The majority of the on hand protocols broadcast information occasionally in a practical method. It increases the count of data transmission as well as the sensed data would be extremely associated. To enhance the energy efficiency, threshold depended on data transmission (reactive) to be projected. The timeline of the reactive data transmission is shown in Fig. 2.

This approach allows the CH to transmit the attributes to its members and the thresholds are listed below:

Hard Threshold **(HT)**: It is a threshold assessment for an attribute that is being sensed. It is the complete rate of the attribute away from that, the node sensing this assessment should control its broadcaster as well as notify to its cluster head.

Soft Threshold **(ST)**: It is a little modification within the rate of the sensed attribute that immunes the node to knob over its transmitter moreover broadcast. The nodes

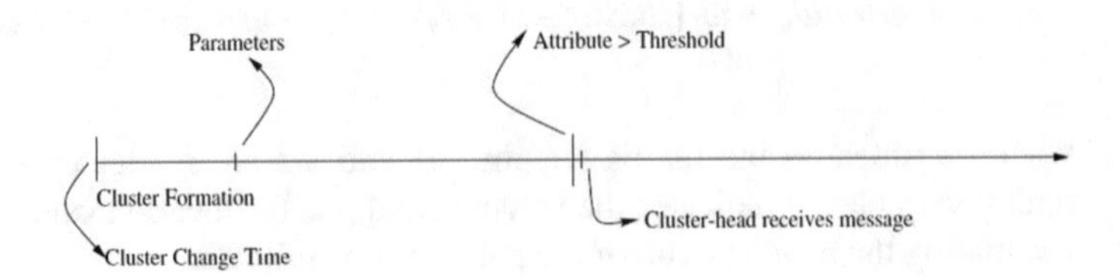

Fig. 2 Time line

sense their environment constantly. The nodes would subsequently broadcast information in the present cluster period, *and* the subsequent circumstances are true:

1. The present rate of the attribute being sensed is higher when compared to the hard threshold.
2. The present rate of the sensed attribute varies from *SV* through a sum equivalent to or superior to the soft threshold. At any time, a node transmits data, *SV* is located equivalent to the in progress value of the sensed attribute.

As a result, the hard threshold minimizes the count of transmissions by allowing the nodes to broadcast alone while the sensed value in the region of significance. The soft threshold additionally minimizes the count of broadcasts through the elimination of every transmission that may contain or else occurred while here is small otherwise no modifications in the sensed attribute on one occasion the rigid threshold.

4 Results and Discussion

The proposed RAP algorithm is implemented in MATLAB and the validation takes place using some metrics. For simulation, 100 sensor nodes are in the area of 100 × 100 m^2. Table 1 depicts the diverse simulation parameters for comparative analysis.

The performance measure throughput indicates the count of packets properly received at the BS. Figure 3 illustrates the comparative results of RAP algorithm with the GSTEB and ACO algorithm. From this figure, it is clearly shown that the RAP algorithm attains maximum throughput than the compared ones. In addition, the inclusion of reactive data transmission acts as a significant task in the improvement of network lifetime. Furthermore, it is verified that the throughput of the RAP algorithm is found to be superior to the existing GSTEB and ACO algorithms.

Table 1 Simulation setup

Parameter	Value
Area (x, y)	100, 100
Base station (x, y)	50, 50 or 50, 150
Nodes (n)	100
Probability (p)	0.1
Initial energy	0.1
Transmitter_energy	$50 * 10^{-9}$
Receiver_energy	$50 * 10^{-9}$
Free space (amplifier)	$10 * 10^{-13}$
Multipath (amplifier)	$0.0013 * 10^{-13}$
Effective data aggregation	$5 * 10^{-9}$
Maximum lifetime	2500
Data packet size	4000

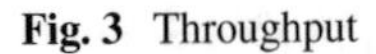
Fig. 3 Throughput

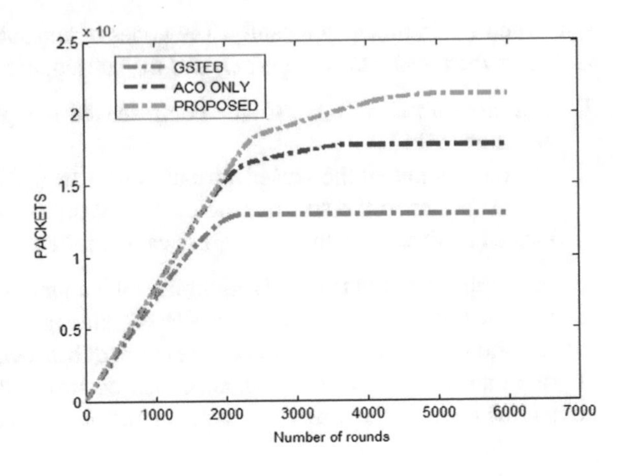

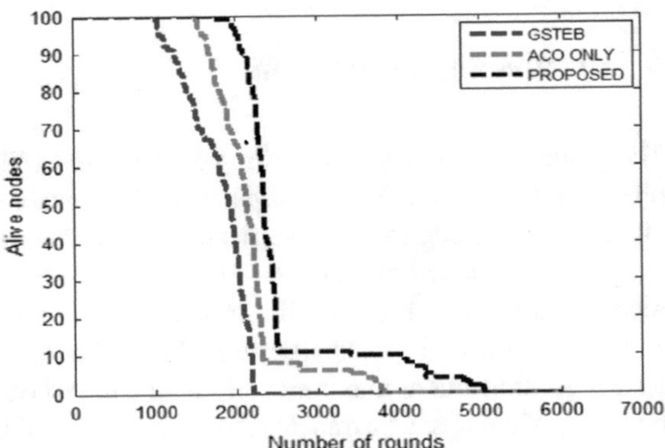

Fig. 4 Analysis of network lifetime

Next, the lifetime of the WSN indicates the time period in between the first as well as last node dies in the WSN. Figure 4 illustrates the comparative analysis of RAP algorithm by means of other methods in terms of lifetime. From the figure, it is clearly depicted that improved lifetime is attained by the RAP algorithm where the network still operates even at 5000 rounds whereas the ACO algorithm makes the network inactive in 3800 rounds itself. It is also apparent that the GSTEB algorithm attains worse performance than the compared methods. On comparing with other algorithms, the RAP algorithm achieved maximum lifetime.

Finally, the residual energy indicates the sum of available energy of all nodes in WSN. Figure 5 provides the comparative results of the proposed RAP algorithm by means of additional algorithms in terms of residual energy. From the figure, it is apparent that the projected RAP algorithm will have maximum residual energy, whereas the GSTEB algorithm achieves minimum residual energy. On the 3000 rounds, the GSTEB algorithm depletes its energy completely whereas the ACO and proposed algorithm have the residual energy of 4 and 7, respectively. From the figures

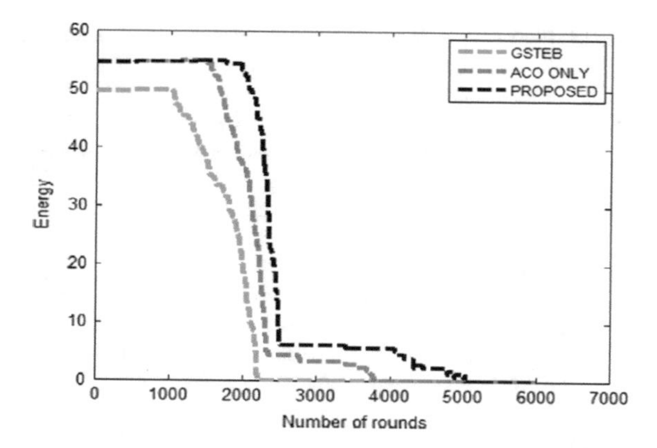

Fig. 5 Analysis of residual energy

and above discussion, it is concluded that the RAP algorithm is found to superior to other algorithms in a significant way.

5 Conclusion

To achieve energy efficiency in WSN, a new RAP protocol is introduced to enhance the network lifetime using the hybridization of ant colony optimization (ACO) as well as particle swarm optimization (PSO) algorithm. To further enhance the energy effectiveness, the proposed RAP algorithm uses a reactive data transmission strategy that is incorporated into the hybridization of ACO and PSO algorithm. In the beginning, the clusters are organized based on the residual energy and then the proposed RAP algorithm will be executed to improvise the inter-cluster data aggregation and reduces the data transmission. The proposed RAP algorithm is implemented in MATLAB and the validation takes place using the subsequent metrics, i.e., stability period, network lifetime, residual energy (average outstanding energy), as well as throughput. The experimental results verified that the RAP algorithm is found superior to other algorithms in a significant way.

References

1. X. Liu, Atypical hierarchical routing protocols for wireless sensor networks: a review. IEEE Sens. J. **15**(10), 5372–5383 (2015)
2. S. Ehsan, B. Hamdaoui, A survey on energy-efficient routing techniques with QoS assurances for wireless multimedia sensor networks. IEEE Commun. Surv. Tutor. **14**(2), 265–278 (2012)
3. O. Younis, M. Krunz, S. Ramasubramanian, Node clustering in wireless sensor networks: recent developments and deployment challenges. IEEE Netw. **20**(3), 20–25 (2006)
4. Y. Mo, B. Wang, W. Liu, L.T. Yang, A sink-oriented layered clustering protocol for wireless sensor networks. Mobile Netw. Appl. **18**(5), 639–650 (2013)

5. M. Elhoseny, X. Yuan, H.K. ElMinir, A.M. Riad, An energy efficient encryption method for secure dynamic WSN, Secur. Commun. Netw. **9**(13), 2024–2031 (2016). https://doi.org/10.1002/sec.1459. (Wiley)
6. M. Elhoseny, X. Yuan, Z. Yu, C. Mao, H. El-Minir, A. Riad, Balancing energy consumption in heterogeneous wireless sensor networks using genetic algorithm. IEEE Commun. Lett. **19**(12), 2194–2197 (2015). https://doi.org/10.1109/lcomm.2014.2381226. (IEEE)
7. Q. Chen, S.S. Kanhere, M. Hassan, Analysis of per-node traffic load in multi-hop wireless sensor networks. IEEE Trans. Wirel. Commun. **8**(2), 958–967 (2009)
8. X. Yuan, M. Elhoseny, H.K. El-Minir, A.M. Riad, A genetic algorithm-based, dynamic clustering method towards improved WSN longevity. J. Netw. Syst. Manag. **25**(1), 21–46 (2017). https://doi.org/10.1007/s10922-016-9379-7. (Springer US)
9. M. Chen, Y. Zhang, Y. Li, M.M. Hassan, A. Alamri, AIWAC: affective interaction through wearable computing and cloud technology. IEEE Wirel. Commun. **22**(1), 20–27 (2015)
10. Y. Zhang, M. Qiu, C.-W. Tsai, M.M. Hassan, A. Alamri, Health CPS: healthcare cyber-physical system assisted by cloud and big data. IEEE Syst. J. **PP**(99), 1–8 (2015)
11. M. Elhoseny, A. Farouk, N. Zhou, M.-M. Wang, S. Abdalla, J. Batle, Dynamic multi-hop clustering in a wireless sensor network: performance improvement. Wirel. Pers. Commun. **95**(4), 3733–3753. https://doi.org/10.1007/s11277-017-4023-8. (Springer US)
12. A. De La Piedra, F. Benitez-Capistros, F. Dominguez, A. Touhafi, Wireless sensor networks for environmental research: a survey on limitations and challenges, in *IEEE EUROCON*, July 2013, pp. 267–274
13. D. Zhang, G. Li, K. Zheng, X. Ming, An energy-balanced routing method based on forward-aware factor for wireless sensor networks. IEEE Trans. Ind. Inform. **10**(1), 766–773 (2014)
14. B. Wang, H.B. Lim, D. Ma, A coverage-aware clustering protocol for wireless sensor networks. Comput. Netw. **56**(5), 1599–1611 (2012)
15. B. Wang, Coverage problems in sensor networks: a survey. ACM Comput. Surv. **43**(4), 32 (2011)
16. M. Elhoseny, A.E. Hassanien, Optimizing cluster head selection in WSN to prolong its existence, in *Dynamic Wireless Sensor Networks*. Studies in Systems, Decision and Control, vol. 165 (Springer, Cham, 2019), pp. 93–111. https://doi.org/10.1007/978-3-319-92807-4_5
17. W. Elsayed, M. Elhoseny, S. Sabbeh, A. Riad, Self-maintenance model for wireless sensor networks. Comput. Electr. Eng. (In Press). https://doi.org/10.1016/j.compeleceng.2017.12.022. Accessed Dec 2017
18. M. Elhoseny, A. Tharwat, A. Farouk, A.E. Hassanien, K-coverage model based on genetic algorithm to extend WSN lifetime. IEEE Sens. Lett. **1**(4), 1–4 (2017). https://doi.org/10.1109/lsens.2017.2724846. (IEEE)
19. B. Singh, D.K. Lobiyal, A novel energy-aware cluster head selection based on particle swarm optimization for wireless sensor networks. Hum.-Centric Comput. Inf. Sci. **2**(1), 1–18 (2012)
20. J. Jin, A. Sridharan, B. Krishnamachari, M. Palaniswami, Handling inelastic traffic in wireless sensor networks. IEEE J. Sel. Areas Commun. **28**(7), 1105–1115 (2010)
21. J. Aweya, Technique for differential timing transfer over packet networks. IEEE Trans. Ind. Inform. **9**(1), 325–336 (2013)
22. J.-D. Tang, M. Cai, Energy-balancing routing algorithm based on LEACH protocol. Comput. Eng. **39**(7), 133–136 (2013)
23. W.R. Heinzelman, A. Chandrakasan, H. Balakrishnan, Energy-efficient communication protocol for wireless microsensor networks, in *Proceedings of the 34th Annual Hawaii International Conference on System Sciences*, Jan. 2000, pp. 1–10
24. O. Younis, S. Fahmy, HEED: a hybrid, energy-efficient, distributed clustering approach for ad hoc sensor networks. IEEE Trans. Mob. Comput. **3**(4), 366–379 (2004)
25. Asaduzzaman, H.Y. Kong, Energy efficient cooperative LEACH protocol for wireless sensor networks. J. Commun. Netw. **12**(4), 358–365 (2010)
26. N. Gautam, J.Y. Pyun, Distance aware intelligent clustering protocol for wireless sensor networks. J. Commun. Netw. **12**(2), 122–129 (2010)

27. A. Manjeshwar, Q.-A. Zeng, D.P. Agrawal, An analytical model for information retrieval in wireless sensor networks using enhanced APTEEN protocol. IEEE Trans. Parallel Distrib. Syst. **13**(12), 1290–1302 (2002)
28. S.D. Muruganathan, D.C.F. Ma, R.I. Bhasin, A.O. Fapojuwo, A centralized energy-efficient routing protocol for wireless sensor networks. IEEE Commun. Mag. **43**(3), S8–S13 (2005)
29. K. Akkaya, M. Younis, A survey on routing protocols for wireless sensor networks. Ad Hoc Netw. **3**(3), 325–349 (2005)
30. X. Gu, J. Yu, D. Yu, G. Wang, Y. Lv, ECDC: An energy and coverage-aware distributed clustering protocol for wireless sensor networks. Comput. Electr. Eng. **40**(2), 384–398 (2014)
31. J. Yu, Y. Qi, G. Wang, X. Gu, A cluster-based routing protocol for wireless sensor networks with nonuniform node distribution. AEU-Int. J. Electron. Commun. **66**(1), 54–61 (2012)
32. J. Yu, Y. Qi, G. Wang, Q. Guo, X. Gu, An energy-aware distributed unequal clustering protocol for wireless sensor networks. Int. J. Distrib. Sens. Netw. **2011** (2011). (Art. no. 202145)
33. A. Chamam, S. Pierre, On the planning of wireless sensor networks: Energy-efficient clustering under the joint routing and coverage constraint. IEEE Trans. Mob. Comput. **8**(8), 1077–1086 (2009)
34. S.K. Singh, M. Singh, D. Singh, A survey of energy-efficient hierarchical cluster-based routing in wireless sensor networks. Int. J. Adv. Netw. Appl. **2**(2), 570–580 (2010)
35. M. Elhoseny, K. Shankar, S.K. Lakshmanaprabu, A. Maseleno, N. Arunkumar, Hybrid optimization with cryptography encryption for medical image security in Internet of Things. Neural Comput. Appl. (2018). https://doi.org/10.1007/s00521-018-3801-x
36. T. Avudaiappan, R. Balasubramanian, S. Sundara Pandiyan, M. Saravanan, S. K. Lakshmanaprabu, K. Shankar, Medical image security using dual encryption with oppositional based optimization algorithm. J. Med. Syst. **42**(11), 1–11 (2018). https://doi.org/10.1007/s10916-018-1053-z
37. S.K. Lakshmanaprabu, K. Shankar, A. Khanna, D. Gupta, J.J. Rodrigues, P.R. Pinheiro, V.H.C. De Albuquerque, Effective features to classify big data using social internet of things. IEEE Access **6**, 24196–24204 (2018)
38. K. Sathesh Kumar, K. Shankar, M. Ilayaraja, M. Rajesh, Sensitive data security in cloud computing aid of different encryption techniques. J. Adv. Res. Dyn. Control. Syst. **9**(18), 2888–2899 (2017)
39. S. Mudundi, H.H. Ali, A new robust genetic algorithm for dynamic cluster formation in wireless sensor networks, in *Proceedings of Wireless and Optical Communications*, Montreal, Quebec, Canada (2007)
40. S. Jin, M. Zhou, A.S. Wu, Sensor network optimization using a genetic algorithm, in *Proceedings of the 7th World Multiconference on Systemics, Cybernetics and Informatics* (2003)
41. M. Elhoseny, A.E. Hassanien, Mobile object tracking in wide environments using WSNs, in *Dynamic Wireless Sensor Networks. Studies*. Systems, Decision and Control, vol. 165. (Springer, Cham, 2009), pp. 3–28. https://doi.org/10.1007/978-3-319-92807-4_1
42. M. Elhoseny, A.E. Hassanien, Expand mobile WSN coverage in harsh environments, in *Dynamic Wireless Sensor Networks*. Studies in Systems, Decision and Control, vol. 165 (Springer, Cham, 2019), pp. 29–52. https://doi.org/10.1007/978-3-319-92807-4_2
43. K. Shankar, P. Eswaran, RGB-based secure share creation in visual cryptography using optimal elliptic curve cryptography technique. J. Circuits Syst. Comput. **25**(11), 1650138 (2016)
44. K. Shankar, P. Eswaran, A secure visual secret share (VSS) creation scheme in visual cryptography using elliptic curve cryptography with optimization technique. Aust. J. Basic Appl. Sci. **9**(36), 150–163 (2015)

Maintaining Consistent Firewalls and Flows (CFF) in Software-Defined Networks

A. Banerjee and D. M. Akbar Hussain

Abstract Software-defined networking (SDN) paradigm brings great flexibility to the network by decoupling control plane from the data plane. However, one of the great security challenges in SDN is to maintain consistency among firewall-rules and actual-flows in the network. The present article proposes one such scheme "consistent firewalls and flows (CFF)" safeguards the network from firewall policy violation and maintains consistency among firewall rules and flow tables. Firewall rule table presented in SDN controller and flow tables present in switches that connect some hosts to the network are treated as critical sections protected by semaphores. We have implemented the CFF framework to demonstrate the efficiency of the proposed scheme and simulation results clearly show benefits of CFF compared to the inbuilt firewall.

Keywords Firewall · Flow table · Security · Semaphore · Software-defined network

1 Introduction

The primary intention behind designing a software-defined network is to implement a centralized control to run various services including packet switching, security mechanisms, network maintenance, etc. The centralized controller, popularly termed as SDN controller is aware of the topology of the whole network. This efficiently decouples control plane from the data plane. OpenFlow protocol [1–5] is mostly used for managing network resources in a cost-effective manner and robust firewalls are required to address security challenges in these types of networks. Firewalls are

A. Banerjee (✉)
Kalyani Government Engineering College, Kalyani, Nadia, West Bengal, India
e-mail: anuradha79bn@gmail.com

D. M. Akbar Hussain
Department of Energy Technology, Section for Power Electronics, Aalborg University, Aalborg, Denmark
e-mail: akh@et.aau.dk

M. Elhoseny and A. K. Singh (eds.), *Smart Network Inspired Paradigm and Approaches in IoT Applications*, https://doi.org/10.1007/978-981-13-8614-5_2

widely deployed security mechanisms used in business and institutions. It sits on the border of a network and examines all incoming and outgoing packets to defend against attacks and unauthorized access. A firewall is built based on the assumption that all elements of the protected network are trusted bodies by themselves and internal traffic need not be monitored and filtered. Possible threats to security systems in SDN arise from dynamic updation of network policy, inconsistency between firewall rules themselves, and inconsistency between firewall rules and flow tables at various switches.

(i) Dynamic updation of network policies—In an open flow network, network states are dynamically updated and configurations are frequently changed. These changes have to be reflected in the firewall as well as flow tables. Ideally, flow tables of associated switches have to be modified before centralized firewall rule table because routing decisions are taken at switches through their flow tables.
(ii) Inconsistency between firewall rules themselves—One firewall rule may contradict the other. For example, (A, B, *, allow) and (A, B, *, deny); here A is source-host-id,and B is destination-host-id. Communication between them on all ports is allowed on the first rule and completely denied in the second rule. This is a contradiction.
(iii) Inconsistency between firewall rules and flow table—Let switch sw1 interface host A to the network. Centralized firewall table contains (A, B, *, deny) whereas the flow table of sw1 contains (A, B). Then this is a contradiction between firewall rules and flow table.

The present article proposes CFF which maintains consistency between firewall rules and flow table. This does not require disturbing the SDN controller for accessing the firewall rule table with the arrival of each packet at an ingress switch. Subparts of firewall rules are copied in relevant switches that easily detect prohibited flows. Moreover, firewall rules are stored in such a manner that makes searching for a particular rule efficient. Rules are stored in increasing order of source-host-id and for each source-host-id, multiple destination-host-ids appear in increasing order . Whenever firewall rule table is going to be modified, some parts of the table are locked in write mode while the other entries are free to be accessed for read or write operation. These techniques greatly enhance the performance of the underlying network.

The present article is organized as follows. We overview related work in Sect. 2. Section 3 discusses CFF with an example. Simulation results appear in Sect. 4 while Sect. 6 concludes the paper.

2 Related Work

Some efforts have been made to tackle security issues in SDN. DDoS attack detection [6], vulnerability assessment [7], saturation attack mitigation [8] etc. are important in this context. However, quite differently from them, our work aims to build efficient firewalls keeping consistency with flow tables of various switches. Floodlight [9] provides us a firewall that stops the undesirable flow of packets at ingress switches depending upon firewall rules "allow" or "deny". However, for each packet flow at the ingress switches, firewall rules present at SDN controller have to be accessed over and over again. This increases the chance of bottleneck at the centralized controller and SDN controller has to be interrupted for every flow, which consumes significant time.

A software extension of FortNOX [10] was proposed for security enforcement in SDN controllers but that did not prove to be very suitable for SDN because FortNOX records rule relations in alias sets, which are unable to track network traffic flows. In [11–14] certain verification tools came up for checking network invariants and policy correctness in OpenFlow but they cannot support automatic and effective violation resolution. Some firewall algorithms and tools have been designed to assist system administrators. Yuan et al. [15] proposed a toolkit to check anomalies or inconsistencies among firewall rules themselves. However, these approaches are not very suitable for software-defined networks.

3 CFF—Explained with an Example

Please consider the network in Fig. 1, consisting of four hosts (A, B, C, D) and seven switches (sw1, sw2, sw3, sw4, sw5, sw6, and sw7). Table 1 specifies firewall rules existing in SDN controller.

In CFF, the controller is equipped with three tables—Firewall-Rules table, Host-Switch table, and Semaphore table. As the name specifies, Firewall-Rules table consists of firewall rules and Host-Switch table consists of host-ids and corresponding switches (switch-id list) that interface a host to the network. For the network shown in Fig. 1, assume that Table 1 shows firewall rules and Table 2 shows Host-Switch table.

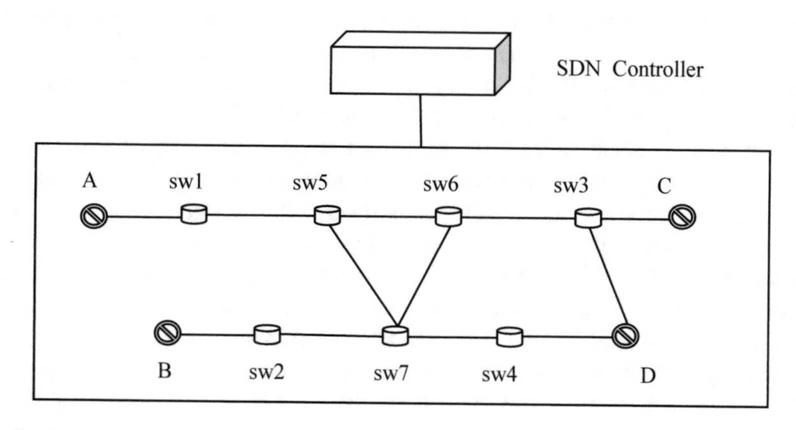

Fig. 1 Example SDN network

Table 1 Firewall Rules table

Source-id	Destination-id	Destination port	Action
A	B	*	Allow
A	C	21, 22	Deny
A	D	21	Allow
B	A	*	Deny
B	C	*	Allow
B	D	23, 25	Deny
C	A	*	Allow
C	B	21	Allow
C	D	*	Deny
D	A	*	Allow
D	B	23	Deny
*	C	*	Deny

Table 2 Host-Switch table

Host-id	Switch-id-list
A	sw1
B	sw2
C	sw3
D	sw3, sw4

Please note that in CFF, firewall rules are arranged in ascending order of source-id and ascending order of destination-id for each source-id.

From the entries of Firewall-Rules table associated with different hosts, two tables named prohibited-flow table and allowed-flow table are constructed corresponding to each switch that interfaces some host to the network. Attributes of prohibited-flow-table are source-id and destination-id whereas the same for allowed-flow-table are source-id, destination-id, and port number. A flow from source H1 to destination H2 will be called prohibited if (H1, H2, *, Deny) is a firewall rule. On the other hand, if firewall rules contain (H1, H2, y, Allow) where y is either * or some port number then (H1, H2) is an allowed flow. Also, if firewall rules contain (H1, H2, y, Deny) where If in the Firewall-Rules table, some port number is mentioned with "Allow" action corresponding to the same source, destination pair, then in allowed-flow table for the same source–destination pair, port number will be positive, whereas for deny action, port number will be negative. For example, in Fig. 1, sw1 interfaces between host A and the network. Therefore, from the firewall rules associated to host A, SDN controller identifies that (A, B), (A, C), and (A, D) are all allowed connections. So, prohibited-flow-table at sw1 is empty and allowed-flow-table is shown in Table 3.

Similarly, prohibited-flow-table and allowed-flow-table for sw2 appear in Tables 4a and 4b, whereas the same tables for sw3 appear in Tables 5a and 5b and for switch sw4, they appear in Tables 6a and 6b.

Table 3 Allowed-flow-table at sw1

Source-id	Destination-id	Port number
A	B	*
A	C	−21, −22
A	D	21

Table 4a Prohibited-flow-table at sw2

Source-id	Destination-id
B	A

Table 4b Allowed-flow-table at sw2

Source-id	Destination-id	Port number
B	C	*
B	D	−23, −25

Table 5a Prohibited-flow-table at sw3

Source-id	Destination-id
C	D
D	C

Table 5b Allowed-flow-table at sw3

Source-id	Destination-id	Port number
C	A	*
C	B	21
D	A	*
D	B	−23

Allowed and prohibited flow tables cannot be independently modified because they are derived from Firewall-Rules table. During modification, flow tables are modified first and if required, the information is propagated to Firewall-Rules table. Reason is that actual routing is performed in nodes based on allowed and prohibited flow tables. If these tables of some host-interface switch are to be modified then no communication session should be able to access those tables. These flow tables are considered to be critical sections and all critical sections are managed using a semaphore. SDN controller maintains a semaphore table too, with attributes Switch-id and Semaphore-id. Semaphore table for the network of Fig. 1, appears in Table 7.

In order to gain access to allowed and prohibited flow tables of sw1, access to semaphore sem1 is required. Similarly, the other entries of Table 7 can be explained.

Table 6a Prohibited-flow-table at sw4

Source-id	Destination-id
D	C

Table 6b Allowed-flow-table at sw4

Source-id	Destination-id	Port number
D	A	*
D	B	−23

Table 7 Semaphore table

Switch-id	Semaphore-id
sw1	sem1
sw2	sem2
sw3	sem3
sw4	sem4

4 Modification in Firewall Rule Table and Flow Table

No direct modification to prohibited and allowed-flow tables can be performed. As far as modification in firewall tables are concerned, violation checks are to be performed. Suppose the firewall rule to be inserted is (α_i, β_i, γ_i, ACT_i) and let, switch (α_i) be switch-id-list of host α_i through which α_i connects to the network. For each switch $s \in$ switch (α_i), new firewall rule has to be compared with prohibited-flow-table(s) and allowed-flow-table(s). Insertion of the new rule in Firewall-Rules table consists of the following steps:

1. If no previous entry with source-destination pair (α_i, β_i) exists in the Firewall-Rules table, then the entry can be directly inserted increasing order of source-id and increasing order of destination-id corresponding to source α_i. Please note that the symbol * denotes the highest among all source-id as well as destination-id.
2. If ACT_i = "Allow", γ_i = * and allowed-flow table of s contains (α_i, β_i, *), then no modification to Firewall-Rules table is required because the new entry is a redundant one.
3. If ACT_i = "Allow", γ_i = * and allowed-flow table of s contains (α_i, β_i, $port_i$), where $port_i$ is the array of port numbers excluding * (irrespective of positive or negative), then a new entry is not inserted in Firewall-Rules table. Just destination port number of the existing entry corresponding to source–destination pair (α_i, β_i) is set to *.
4. If ACT_i = "Allow" γ_i ! = * and allowed-flow table of s contains (α_i, β_i, *), then no modification needs to be performed.
5. If ACT_i = "Allow" γ_i ! = * and allowed-flow table of s contains (α_i, β_i, $port_i$), where $port_i$ is the array of port numbers excluding *, then new $port_i = port_i \cup \gamma_i$.
6. If ACT_i = "Deny", and source–destination pair (α_i, β_i) is present in prohibited-flow-table of s, then no modification is performed.
7. If ACT_i = "Deny", γ_i = * and source-destination pair (α_i, β_i) is present in allowed-flow-table of s, then a violation is reported.
8. If ACT_i = "Deny" γ_i ! = * (i.e., some set of valid port numbers) and source-destination pair (α_i, β_i) is present in allowed-flow-table of s with $port_i$ = *, then

$(\alpha_i, \beta_i, *)$ is replaced by $(\alpha_i, \beta_i, -x1, -x2, \ldots, -xj)$ where, for all $1 \leq j \leq k$, $xj \in \gamma_i$. This modification is propagated to firewall rule table of the controller.

9. If ACT_i = "Deny" γ_i ! = * (i.e., some set of specific port numbers) and source-destination pair (α_i, β_i) is present in allowed-flow-table of s with $port_i$! = * (i.e., some set of specific port numbers), then new $port_i = port_i - \gamma_i$.. If $port_i = \{\emptyset\}$ then $(\alpha_i, \beta_i, port_i)$ is deleted from allowed-flow-table(s) and (α_i, β_i) is inserted in prohibited-flow-table(s). This modification is propagated to Firewall-Rules table of the controller.

Example

Suppose in Firewall-Rules in Table 1 the following record needs to be inserted

D	C	22	Deny

Since source–destination pair (D, C) is not already there in Firewall-Rules table, (D, C, 22, Deny) should precede (*, C, *,Deny). Accordingly, prohibited and allowed rule tables of the switches associated to host D have to be modified. D connects to the network through sw3 and sw4. So, access to sem3 and sem4 need to be acquired before the task. After the access is gained, new prohibited flow table of sw3 is shown in Table 8. Prohibited flow table of sw4 will become empty. New allowed-flow tables of sw3 and sw4 are shown in Tables 9 and 10.

Table 8 Prohibited flow table At sw3

Source-host-id	Destination-host-id
C	D

Table 9 Allowed-flow-table at sw3

Source-id	Destination-id	Port number
C	A	*
C	B	21
D	A	*
D	B	−23
D	C	−22

Table 10 Allowed-flow-table at sw4

Source-id	Destination-id	Port number
D	A	*
D	B	−23
D	C	−22

5 Implementation

CFF is implemented on a network with 10 hosts and 20 Ethernet switches. Simulation metrics are firewall rule update time (in microsecond) and per packet inspection time (in microsecond). Used communication protocol is OpenFlow. Performance of CFF is compared with the built-in firewall. The graphical comparison appears in Figs. 2 and 3.

CFF focuses on the fact that the switch interfacing a host to a network, is only aware of the actual source of communication. Therefore, firewall rules monitoring all kinds of communications from that particular source are placed by CFF in the switch that interfaces the host to the network. In CFF, policy updates are fired on allowed and prohibited flow tables of those interfacing switches. Validations are performed in switches only and modifications are propagated to firewall rules table if necessary. Otherwise, the centralized controller of SDN is not disturbed. This is the novelty of CFF that greatly reduces firewall rule update time and per packet inspection time.

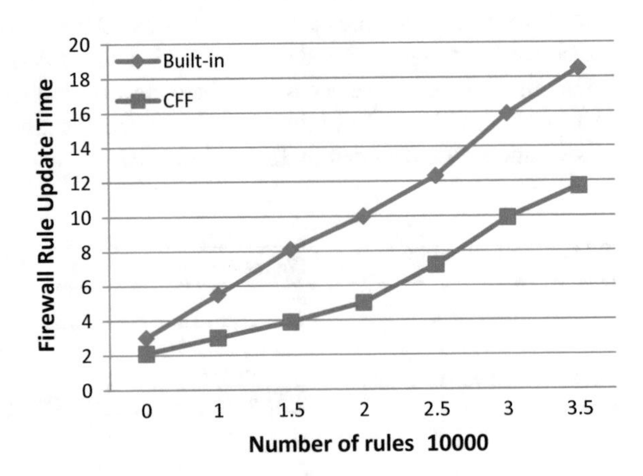

Fig. 2 Firewall rule update time in microsecond versus the number of rules

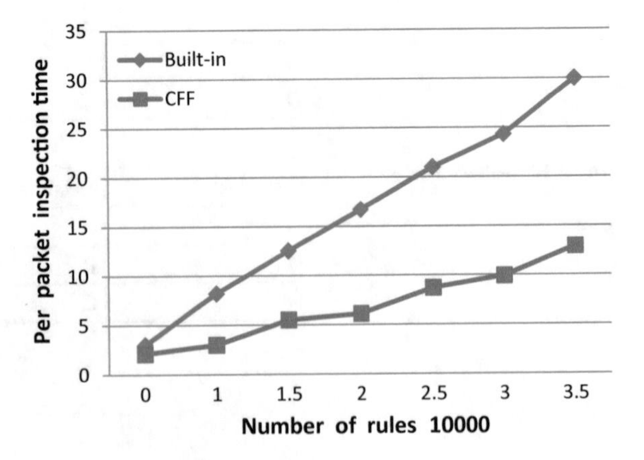

Fig. 3 Per packet inspection time in microsecond versus the number of rules

6 Conclusion

CFF is a technique to maintain consistency between the table of firewall rules and flow tables. Also, it safeguards the network from firewall policy violation. Allowed and prohibited flow tables are maintained as critical sections protected by semaphores. SDN controller uses a semaphore table to store information about them corresponding to each host-interfacing switch in the network. Instead of accessing the firewall rules every now and then, flow tables at switches are modified and modifications are reflected in firewall rules table only if required. This tremendously reduces firewall rule update time and per packet inspection time.

References

1. N. McKeown, T. Anderson, H. Balakrishnan, G. Parulkar, L. Peterson, J. Rexford, S. Shenker, J. Turner, Openflow: enabling innovation in campus networks. ACM SIGCOMM Comput. Commun. Rev. (2008)
2. S. Jadala, S. Pelluri, Energy optimization at cloud data centers using SDN. Int. J. Eng. Trends Technol. Special issue (2017, April)
3. K. Xie, S. Hao, M. Ma, E3MC: Improving Energy Efficiency via Elastic Multi-controller SDN in Data Center Networks, IEEE Access, November 2016
4. M. Rifai et. al., MINNIE: an SDN world with few compressed forwarding rules. Comput. Netw. 121 (2017, July)
5. H. Hu, W. Han, G.J. Ahn, Z. Zhao, FLOWGUARD: building robust firewalls for software-defined networks, in *ACM SIGCOMM Workshop on Hot Topics in Software Defined Networking (HotSDN) 2014*
6. B. Braga, M. Mota, P. Passito et al., Lightweight DDoS flooding attack detection using NOX/OpenFlow, in *LCN'10*
7. D. Kreutz, F. Ramos, P. Verissimo, Towards secure and dependable software-defined networks, in *HotSDN'13*
8. S. Shin, V. Yegneswaran, P. Porras, G. Gu, Avant-guard: scalable and vigilant switch flow management in software-defined networks, in *CCS'13*
9. Floodlight: Open SDN Controller. http://www.projectfloodlight.org
10. P. Porras, S. Shin, V. Yegneswaran, M. Fong, M. Tyson, G. Gu, A security enforcement kernel for openflow networks, in *HotSDN'12*
11. P. Kazemian, M. Chang, H. Zeng, G. Varghese, N. McKeown, S. Whyte, Real time network policy checking using header space analysis, in NSDI'13
12. P. Kazemian, G. Varghese, N. McKeown, Header space analysis: static checking for networks, in *NSDI'12*
13. A. Khurshid, X. Zou, W. Zhou, M. Caesar, P.B. Godfrey, Veriflow: verifying network-wide invariants in real time, in *NSDI'13*
14. H. Mai, A. Khurshid, R. Agarwal, M. Caesar, P. Godfrey, S.T. King, Debugging the data plane with anteater, in *SIGCOMM'11*
15. L. Yuan, H. Chen, J. Mai, C. Chuah, Z. Su, P. Mohapatra, C. Davis, Fireman: a toolkit for firewall modeling and analysis, in *2006 IEEE Symposium on Security and Privacy*
16. Frenetic: A Family of Network Programming Languages. http://frenetic-lang.org/
17. Header Space Library. https://bitbucket.org/peymank/hassel-public
18. E. Al-Shaer, H. Hamed, Discovery of policy anomalies in distributed firewalls, in *INFOCOM'04*
19. G. Bianchi, M. Bonola, A. Capone, C. Cascone, OpenState: programming platform-independent stateful openflow applications inside the switch. ACM SIGCOMM Comput. Commun. Rev. (2014)

20. S.K. Fayazbakhsh, L. Chiang, V. Sekar, M. Yu, J.C. Mogul, Enforcing network-wide policies in the presence of dynamic middlebox actions using flowtags, in *NSDI'14*
21. H. Hu, G.-J. Ahn, K. Kulkarni, FAME: a firewall anomaly management environment, in *SafeConfig'10*
22. H. Hu, G.-J. Ahn, K. Kulkarni, Detecting and resolving firewall policy anomalies. IEEE Trans. Dependable Secur. Comput. **9**(3), 318–331 (2012)
23. S. Ioannidis, A.D. Keromytis, S.M. Bellovin, J.M. Smith, Implementing a distributed firewall, in *CCS'00*
24. S.A. Mehdi, J. Khalid, S.A. Khayam, Revisiting traffic anomaly detection using software defined networking, in *RAID'11*
25. C. Monsanto, J. Reich, N. Foster, J. Rexford, D. Walker, Composing software-defined networks, in *NSDI'13*
26. M. Reitblatt, N. Foster, J. Rexford, C. Schlesinger, D. Walker, Abstractions for network update, in *SIGCOMM'12*
27. E.E. Schultz, A framework for understanding and predicting insider attacks. Comput. Secur. **21**(6), 526–531 (2002)
28. S. Shirali-Shahreza, Y. Ganjali, Flexam: flexible sampling extension for monitoring and security applications in openflow, in *HotSDN'13*
29. R. Stoenescu, M. Popovici, L. Negreanu, C. Raiciu, Symnet: static checking for stateful networks, in *HotMiddlebox'13*
30. J. Wang, Y. Wang, H. Hu, Q. Sun, H. Shi, L. Zeng, Towards a security-enhanced firewall application for openflow networks, in *Cyberspace Safety and Security* (2013)

Energy-Efficient Broadcasting of Route-Request Packets (E^2BR^2) in Ad Hoc Networks

Anuradha Banerjee and Subhankar Shosh

Abstract Broadcasting is a very important operation in ad hoc networks. It helps in discovering routes to unknown destinations and repair links in case of breakage. As far as the broadcasting of route-requests (RREQ) is concerned, flooding is the only method in which, each node sends a broadcast packet to all of its one-hop downlink neighbors. E^2BR^2 introduces a very novel observation that broadcasting of RREQ is much different from the broadcasting of data packets, and therefore, power conserving requirements and methodologies are different; especially when a recent location of the destination is known. E^2BR^2 instructs each node to keep track of its two-hop downlink neighbors to take advantage of topological redundancies wherever present. Simulation results show that the proposed technique greatly improves network throughput, significantly saves rebroadcast reducing energy consumption and delay.

Keywords Ad hoc networks · Broadcasting · Energy efficiency · Redundancy · Topology

1 Introduction

An ad hoc network finds extensive applications in a natural disaster, military operations, etc. It does not require any fixed infrastructure or centralized administration [1–7]. Nodes move freely with random velocity and direction. They may act as end-points or routers to forward packets in a multi-hop environment. The role is that of an endpoint when they either initiate communication or specified as the destination by some node, and that of a router, when they are elected by the destination to bridge the gap between generating and receiving sites of a broadcast packet. Broadcast-

A. Banerjee (✉)
Kalyani Government Engineering College, Kalyani, West Bengal, India
e-mail: anuradha79bn@gmail.com

S. Shosh
Regent Education and Research Foundation, Kolkata, India
e-mail: redhatsubha@gmail.com

M. Elhoseny and A. K. Singh (eds.), *Smart Network Inspired Paradigm and Approaches in IoT Applications*, https://doi.org/10.1007/978-981-13-8614-5_3

ing of route-requests (RREQ) is very crucial in ad hoc networks because it is the most primitive operation required for unicasting and multicasting, that is, when one particular node or a predefined subset of network nodes, are to be discovered [3, 5, 8]. To the best of authors' knowledge, the existing literature on ad hoc networks is completely silent about the broadcasting of RREQ packets. The differences between broadcasting of RREQ and data packets are completely unexplored.

To the best of the authors knowledge, no specific algorithm for energy-efficient broadcasting of route-request packets, exist in the literature on ad hoc networks. Proposed work is the first to focus on the fact that broadcasting of data is much different from the broadcasting of route-request packets. Broadcasting of data intends to forward the broadcast packet to all nodes in the network whereas flooding is utilized in all ad hoc routing protocols which forward the route-request packet to all nodes in the network simply to reach one particular node, i.e., the destination. E^2BR^2 aims at reducing the number of route-request packets injected into the network. Two-hop neighborhood information is maintained at each node. Instead of broadcasting route-request within its entire neighborhood, a node in E^2BR^2 embedded routing protocol (the routing protocol can be any standard protocol in ad hoc networks), finds out minimum energy path to each two-hop neighbor. In this way, some of its one-hop neighbors can be eliminated from consideration, that is, route-request packet won't be sent to them. Simulation is performed using ns-2 simulator. Results shown in Sect. 5 are very encouraging. They show significant improvement in favor of the proposed scheme.

In Sect. 2, we review the previous work in broadcasting as a whole. Section 3 clearly illustrates why broadcasting of RREQ is different from the broadcasting of data packets. E^2BR^2 is explained along with the network model and example, in Sect. 4. Given the discovered QoS classes, Sect. 5 presents the simulation results while Sect. 6 concludes the paper.

2 Related Work

The simplest method of broadcasting that is used for forwarding of both RREQ and data packets, is blind flooding where each router forwards the broadcast packets to all of its one-hop neighbors without considering redundancy is topology or essential downlink neighbors in case of RREQ. Although its packet delivery rate is high but that is achieved at a huge cost of messages (popularly termed as broadcast storm problem [5, 9, 10] unnecessarily consuming the energy of nodes and increasing the delay. To mitigate this problem, several schemes have been proposed for the broadcasting of data [9, 11–17] but none are there focussing on RREQs.

A probability-based broadcast scheme is proposed in [11], where a node rebroadcasts provided its downlink neighbor density is high. Otherwise, it simply drops the broadcast packet. Downlink neighbor density of a node is calculated as {(total number of downlink neighbors)/($\pi \times$ (radio-range)^2)}. A dynamic probabilistic broadcast scheme is proposed in [12], where a combination of probabilistic- and counter-based approach is applied. A local packet counter is maintained at each node that contributes the to adjustment of rebroadcast probability. The job of local packet counter is to keep a record of duplicate copies of the same packet. If its values are less than a predefined threshold and one-hop downlink neighbor density is high, then only the corresponding router forwards the packet. A color-based broadcast scheme appears in [9] where a color field is associated to every broadcast message. Colors actually differentiated between duplicate copies of the same message. Due to topological redundancy, a node may overhear the same broadcast message more than once. Colors associated with these messages are generally different because whenever a router rebroadcasts a message, it assigns a new color to the packet. If the number of colors of that message is less than a predefined threshold, then the routers rebroadcast; otherwise, it drops the message. Certain counter-based schemes are proposed in [13, 14] where broadcast probability is set to 0.65. The authors have claimed in the simulation results section that this broadcast probability (0.65) produces good packet delivery ratio decreasing message cost. A reliable broadcast scheme appears in [15] where an effective one-hop neighbor set of a router is determined from their distances from a broadcast source. Those which are farther from the broadcast source are given preference.

Directional gossip (DG [16]) is a protocol that instructs nodes to assign weights to all of its one-hop neighbors based on their connectivity. Connectivity is measured in terms of residual energy and history of survival of the link. If residual energy is high and the link between those two nodes survived for a long time in earlier sessions, then that is expected to live long in the current session too. The broadcast packet is forwarded to the neighbors with low weight, because, weight is defined as (1/connectivity). Location-based broadcast is another important scheme where each node that has received a broadcast message, computes additional coverage area of its one-hop downlink neighbors compared to the broadcast source. The neighbors that produce high additional coverage area, are forwarded the broadcast packet.

E^2BR^2 can be implemented along with any reactive routing protocol applicable to ad hoc networks. Here, we consider two state-of-the-art reactive protocols—AODV (ad hoc on-demand distance vector) and ABR (associativity based routing). Flooding is used for route discovery in all reactive protocols, so in AODV and ABR. Among all the paths through which RREQs arrive at the destination, the one with least number of hops is elected for communication in AODV. On the other hand, ABR is concerned with the stability of communication links. An associativity table is maintained at each node and nodes exchange periodic beacons. Associativity table records estimated stability of connection from a node to its one-hop neighbors. A link with higher associativity tick is termed as stable. In the simulation section, ordinary version (where routes are discovered through blind flooding) of these two popular protocols are compared with E^2BR^2 embedded versions of the same protocols. Simulation results show emphatic improvements in favor of E^2BR^2.

3 Why Broadcasting of RREQ Is Special?

Unlike data, broadcasting of RREQ packets do not require pair-wise acknowledgments (hereinafter shall be referred to as ack). For example, consider that a broadcast source n_s initiates an RREQ packet targeting destination n_d and eventually the RREQ arrives at a router n_r. Blind flooding says n_d to broadcast the packet to all of its one-hop downlink neighbors, say n_p, n_q, n_l, and n_v. Receiving this, all four of n_p, n_q, n_l, and n_v forward the same RREQ within their own respective radio-ranges. They do not need to send an ack to n_r. Only after the RREQs arrive at the destination n_d, n_d elects one or more (at most three, depending upon the underlying routing protocol. It is one in case of AODV and ABR) route from n_s to itself and embeds the optimal route information in RREP packet which flows back from destination to source. Would it have been broadcasting of data, then immediately after receiving the broadcast packet, all four of n_p, n_q, n_l, and n_v had to send ack to n_r, before forwarding the broadcast packet within their own radio-circles.

4 Proposed Work E^2BR^2

In this section, an energy-efficient RREQ broadcast method is proposed, that takes into account the advantages of both topology control and power management. Redundancies are eliminated after every two hops in order to achieve energy preservation and controlled flooding instead of the blind one.

4.1 Network Model

Underlying ad hoc network is modeled as a graph G = (V, E), where V is the set of vertices and E is the set of edges. Certain heterogeneous nodes are deployed in the network having differences in maximum battery power, radio-range, antenna capacity, and maximum speed. Each node is equipped with a unique identification number and directional antenna that can select a set of nodes to which a broadcast message can be sent.

Each node n_i transmits hello message at regular intervals throughout its radio-circle. Components of this message are as follows:

(i) sender-id
(ii) location
(iii) radio-range
(iv) maximum transmission power
(v) timestamp

As the name reveals, sender-id is a unique identification number of the sender, location is the present geographical position as an ordered pair of latitude and longitude.

Radio-range is the radius of the radio-circle. If a node n_m is present within the radio-circle of n_i, then n_m can directly receive messages from n_i. Maximum transmission power depends upon the antenna, whereas timestamp specifies the current time.

Nodes residing within radio-range of n_i reply with ack consisting of the following fields:

(i) sender-id
(ii) location
(iii) radio-range
(iv) maximum transmission power
(v) minimum receive power
(vi) single hop neighbor list (SNL)
(vii) timestamp

Significance of attributes common to both hello and ack are already mentioned earlier in this section. Minimum receive power is once again an antenna-dependent attribute. SNL containing the following attributes:

(i) downlink neighbor id
(ii) location
(iii) minimum receive power

Components of RREQ messages are as follows:

(i) sender-id
(ii) destination-id
(iii) cur-hopcount
(iv) broadcast information list (BIL)
(v) timestamp of initiation

BIL may be empty or may contain the following information:

(i) relevant downlink neighbor id
(ii) broadcast path
(iii) energy cost

cur-hopcount denotes the number of hops till the current router. It is incremented by one, whenever a router broadcasts an RREQ. cur-hopcount must be less than or equal to hop count where hop count is the maximum possible number of hops a packet can travel in an ad hoc network. Whenever a broadcast source injects a new RREQ in the network, cur-hopcount is initialized to zero so that immediately from the next hop, two-hop neighborhood information of a node can be applied in order to detect topological redundancies.

As soon as RREQ arrives at a router n_v, it will first check the value of cur-hopcount. If it is zero or even, then energy-efficient routes to two-hop neighbors (all or a relevant subset provided a recent location of the destination is known) are discovered using the route selection techniques in E^2BR^2. On the other hand, if cur-hopcount is odd, then some RREQ forwarding instruction must have arrived from the predecessor in the form of BIL. The term relevant downlink neighbor is clarified in Sect. 4.2.1.

Please note that each node maintains a cache memory CM of last M number of destinations discovered so far. It contains the following fields:

(i) destination-id
(ii) location
(iii) timestamp

Significance of attributes of CM is explained earlier in this section.

4.2 Broadcast Route Selection for RREQ Packets

4.2.1 When No Recent Location of the Destination Is Known

Please consider Fig. 1 where RREQ packet has just arrived at n_v from n_s.

Node n_v initiates energy optimization within its two-hop neighborhood. So, it is marked in blue in Fig. 1. Its one-hop downlink neighbors are green and two-hop neighbors are marked in brick red. This color convention will be followed throughout the article.

Content of the RREQ packet received by n_v, is shown in Fig. 2, where sender-id and destination-id are n_s and n_d, respectively; cur-hopcount is 0 because the RREQ has just been initiated. For the same reason BIL is empty; some router receiving the RREQ with cur-hopcount being zero or even, will change BIL for the next hop downlink neighbor(s). t_s is the timestamp of initiation of the RREQ.

Immediately after receiving the RREQ, n_v checks whether the destination n_d resides within its two-hop neighborhood. If n_d is a direct neighbor of n_v, then n_v

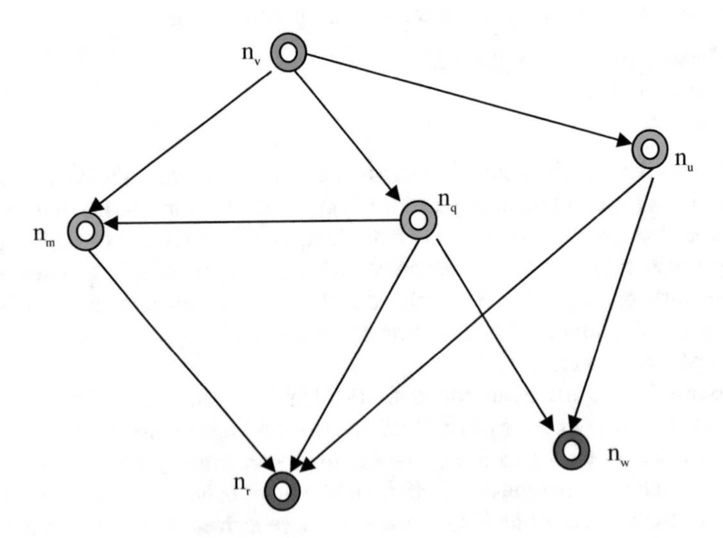

Fig. 1 Network connectivity graph of G

n_s	n_d	0	{∅}	t_s

Fig. 2 RREQ packet that arrived at n_v

directly sends the RREQ to it. On the other hand if n_d is a two-hop neighbor of n_v, then n_v computes most energy-efficient route from itself to n_d. But if it happens that none of the one-hop or two-hop neighbors of n_v is n_d, then n_d targets only its two-hop neighbors and finds energy-efficient routes to them. The one-hop neighbors that reside in those energy-efficient paths are considered as relevant downlink neighbor while other one-hop neighbors are simply discarded, because the ultimate aim in route discovery is to drive the RREQ packets toward the destination (through as much energy-efficient paths as possible) and not to all nodes in the network as in broadcasting of data.

Assuming that $p_{max}(v)$ and $p_{recv}(v)$ are the maximum transmission power and minimum receive power of n_v, respectively, overall broadcast cost $OVC_G(v)$ in two-hop neighborhood of n_v in network G is formulated in (1).

$$\begin{aligned} OVC_G(v) = {} & f_v(q) + f_v(m) + f_v(u) + f_q(m) + f_m(r) + f_q(r) \\ & + f_u(r) + f_u(v) + f_q(w) \end{aligned} \tag{1}$$

Here

$$f_a(b) = \{p_{recv}(b) \times (dist_{ab})^j\}/\beta \tag{2}$$

$dist_{ab}$ is the most recent cartesian distance of n_b from n_a, as shown in (3).

$$dist_{ab} = \sqrt{\left\{(x_a - x_b)^2 + (y_a - y_b)^2\right\}} \tag{3}$$

$j = 2, 3$ or 4 depending upon the medium of communication. β is a constant. x_a, x_b, y_a, and y_b denote x coordinates of n_a and n_b followed by y coordinates of those two nodes, respectively.

SNL of different nodes in Fig. 1 appear in Tables 1, 2, 3, and 4.

Please note that, targeted two-hop neighbors of n_v are n_r and n_w. Possible communication routes from n_v to n_r are as follows :

Table 1 SNL of n_v

Downlink neighbor id	Recent location	Minimum receive power
n_m	(x_m, y_m)	$p_{recv}(m)$
n_q	(x_q, y_q)	$p_{recv}(q)$
n_u	(x_u, y_u)	$p_{recv}(u)$

Table 2 SNL of n_m

Downlink neighbor id	Recent location	Minimum receive power
n_r	(x_r, y_r)	$p_{recv}(r)$

Table 3 SNL of n_u

Downlink neighbor id	Recent location	Minimum receive power
n_r	(x_r, y_r)	$p_{recv}(r)$
n_w	(x_w, y_w)	$p_{recv}(w)$

Table 4 SNL of n_q

Downlink neighbor id	Recent location	Minimum receive power
n_r	(x_r, y_r)	$p_{recv}(r)$
n_w	(x_w, y_w)	$p_{recv}(w)$

(i) $n_v \rightarrow n_m \rightarrow n_r$
(ii) $n_v \rightarrow n_q \rightarrow n_m \rightarrow n_r$
(iii) $n_v \rightarrow n_q \rightarrow n_r$
(iv) $n_v \rightarrow n_u \rightarrow n_r$

Similarly, possible routes n_v to n_w are as follows:

(i) $n_v \rightarrow n_q \rightarrow n_w$
(ii) $n_v \rightarrow n_u \rightarrow n_w$

We shall discover energy-efficient routes to both n_r and n_w, one after another.

Let $e\_cost_{v,m}(\{\emptyset\})$ and $e\_cost_{v,m}(\{q\})$ denote energy cost of communication from n_v to n_m with empty set of routers and the same with router n_q, respectively. Expressions for these two are formulated in (4) and (5), respectively.

$$e\_cost_{v,m}(\{\emptyset\}) = f_v(m) \quad (4)$$

$$e\_cost_{v,m}(\{q\}) = f_v(q) + f_q(m) \quad (5)$$

If $e\_cost_{v,m}(\{\emptyset\}) < e\_cost_{v,m}(\{q\})$, then broadcast path from n_v to n_m is the direct one $n_v \rightarrow n_m$. The associated energy cost will be $e\_cost_{v,m}(\{\emptyset\})$. Otherwise broadcast path from n_v to n_m will be $n_v \rightarrow n_q \rightarrow n_m$ with the corresponding energy cost being $e\_cost_{v,m}(\{q\})$. We are particularly interested in computing the most efficient route to n_m because there exists a path from n_m to n_r which is one of our targeted two-hop downlink neighbor of n_v.

$$e\_cost_{v,r}(\{m\}) = f_v(m) + f_m(r) \quad (6)$$

$$e\_cost_{v,r}(\{q, m\}) = f_v(q) + f_q(m) + f_m(r) \quad (7)$$

$$e\_cost_{v,r}(\{q\}) = f_v(q) + f_q(r) \quad (8)$$

$$e\_cost_{v,r}(\{u\}) = f_v(u) + f_u(r) \quad (9)$$

Let $\exists$ k $\in$ {{m},{q, m}, {q},{u}} s.t. condition (10) holds good.

$$e\_cost_{v,r}(\{k\}) \leq_{\forall i \in \{\{m\},\{q,m\},\{q\},\{u\}\}} e\_cost_{v,r}(\{i\}) \quad (10)$$

Then applicable broadcast path from n_v to n_r in E^2BR^2 will consist of all intermediate nodes in k and the corresponding energy cost is $e\_cost_{v,r}(\{k\})$.

Similarly, if $e\_cost_{v,w}(\{q\}) < e\_cost_{v,w}(\{u\})$, then applicable broadcast path from n_v to n_w in E^2BR^2 will be $n_v \rightarrow n_q \rightarrow n_w$ with the associated broadcast cost being $e\_cost_{v,w}(\{q\})$. Otherwise, it will be $n_v \rightarrow n_u \rightarrow n_w$ while the broadcast cost is $e\_cost_{v,w}(\{u\})$.

In case if there is a tie between energy consumptions belonging to various paths, the one that produces lesser delay, will be elected as optimal. For example, please assume that

Conditions (11) and (12) hold good.

$$\left(e\_cost_{v,w}(\{q\}) = e\_cost_{v,w}(\{u\})\right) \quad (11)$$

$$\left(delay_{v,w}(\{q\}) < delay_{v,w}(\{u\})\right) \quad (12)$$

$delay_{v,w}(\{q\})$ specifies composite delay in the path $n_v \rightarrow n_q \rightarrow n_w$.

Without any loss of generality, let us assume that least energy consuming paths from n_v to n_r is $n_v \rightarrow n_q \rightarrow n_r$ and the same from n_v to n_w is $n_v \rightarrow n_q \rightarrow n_w$. Then, connectivity in reduced graph G′ is shown in Fig. 3.

$$OVC_{G'}(v) = f_v(q) + f_q(r) + f_q(r) \quad (13)$$

Improvement produced by G′ over G in terms of energy is denoted as $I_{G,G'}(v)$ and defined in (14).

$$I_{G,G'}(v) = (1 - OVC_{G'}(v)/OVC_G(v)) \times 100\% \quad (14)$$

Here we find from Fig. 3 that there exists only one relevant one-hop downlink neighbor of n_v, which is n_q. Therefore, n_v sends BIL to n_q only (as shown in Table 5); not to n_m or n_u.

Among one-hop downlink neighbors n_m and n_u are completely discarded because they are irrelevant as far as two-hop downlink neighbors of n_v are concerned. RREQ forwarded by n_v to n_q is shown in Fig. 4.

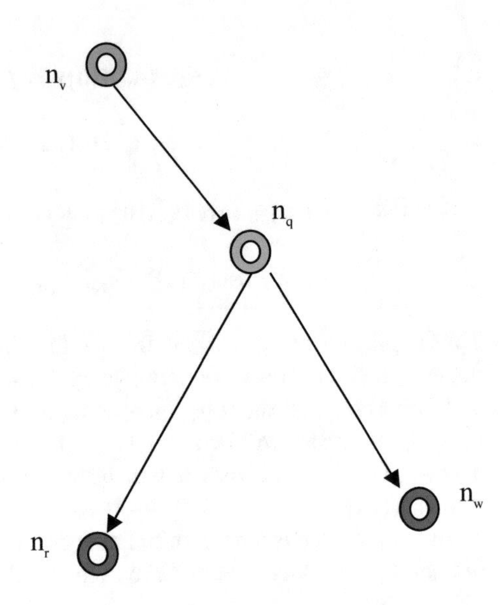

Fig. 3 Connectivity graph in G′

Table 5 BIL of n_v

Single hop downlink neighbor id	Broadcast path	Energy cost
n_q	(i) $n_v \rightarrow n_q \rightarrow n_r$	$e\_cost_{v,r}(\{q\})$
	(ii) $n_v \rightarrow n_q \rightarrow n_w$	$e\_cost_{v,w}(\{q\})$

n_s	n_d	1	BIL of table 5	t_s

Fig. 4 RREQ forwarded by n_v to n_q

n_s	n_d	2	{∅}	t_s

Fig. 5 RREQ forwarded by n_q to n_r and n_w

Receiving this RREQ, n_q forwards that to n_r and n_w (with some modifications) through the paths mentioned in BIL of received RREQ. Figure 5 produces the RREQ forwarded by n_q.

n_r and n_w will apply two-hop neighborhood information to generate their own BIL to respective one-hop relevant downlink neighbors.

4.2.2 A Recent Location of the Destination Is Known

Please consider Fig. 1 where RREQ packet has just arrived at n_v from n_s. For each RREQ packet, a time-to-live (TTL) attribute can be computed as follows:

$$TTL = H \times \left(R_{min} + R_{max}\right)/(2 \times vs) \quad (15)$$

where H is maximum possible number of hops in the network. R_{min} and R_{max} are minimum and maximum possible radio-ranges in the network whereas vs is the velocity of the wireless signal. The formulation in (15) is based on the fact that in every hop we are assuming that the signal covers the approximate distance $((R_{min} + R_{max})/2)$.

Let router n_v last interacted with destination n_d at time t', i.e., CM of n_v contains the entry (n_d, t') and a cache hit took place at current time t. Lifetime left τ_l for the current RREQ appears in (16).

$$\tau_l = TTL-(t-t_s) \tag{16}$$

Within the time period $(t - t' + \tau_l)$, the position of destination is bound by a location circle LC with center (x_d, y_d) and radius $\{vel_{max}\ (t - t' + \tau_l)\}$ where vel_{max} is the maximum possible velocity of any node in the network. Assuming that a rectangle ABCD circumscribes the circle LC, then coordinates of A, B, C, and D are shown below:

(i) A (left top coordinate position LT_x, LT_y)
(ii) B (right top coordinate position RT_x, RT_y)
(iii) C (left bottom coordinate position LB_x, LB_y)
(iv) D (right bottom coordinate position RB_x, RB_y)

where

$$LT_x = x_d - \{vel_{max}(t - t' + \tau_l)\} \tag{17}$$

$$LT_y = y_d + \{vel_{max}(t - t' + \tau_l)\} \tag{18}$$

$$LT_x = x_d + \{vel_{max}(t - t' + \tau_l)\} \tag{19}$$

$$RT_y = LT_y \tag{20}$$

$$LB_x = LT_x \tag{21}$$

$$LB_y = y_d - \{vel_{max}(t - t' + \tau_l)\} \tag{22}$$

$$RB_x = RT_x \tag{23}$$

$$RB_y = LB_y \tag{24}$$

Assume that from source node n_s to current router n_v, number of hops is λ. Therefore, the remaining number of hops from n_q is $(H - \lambda - 1)$. If it is impossible to reach all four vertices A, B, C, and D from any $n_i \in \{n_r, n_w\}$, then n_v instructs n_q to discard n_i. The condition is expressed mathematically in (25).

$$\forall_{Z \in \{\{A\},\{B\},\{C\},\{D\}\}} DT_Z(i) > \{(H - \lambda - 1) \times (R_{min} + R_{max})\}/2 \quad (25)$$

$DT_Z(i)$ is cartesian distance from n_i to vertex Z of the rectangle circumscribing LC of n_d.

If condition (25) is false, then energy-efficient routes are discovered from n_v to n_i using the procedure in Sect. 4.2.1. Please note that in E^2BR^2, only those nodes are discarded from which it is really impossible to track the destination within residual number of hops left. This is quite practical and applicable for RREQs compared to data packets, because during the broadcasting of data packets, all nodes in the network are destinations. RREQs in unicasting has only one destination and the same injected for multicasting, has a small subset of network nodes as destinations. Discarding irrelevant two hop downlink neighbors promote energy saving by avoiding unnecessary transmission to those two-hop neighbors from which destination cannot be reached within remaining hop count.

5 Illustration with an Example

Consider network connectivity graph of Fig. 6.

Let n_v knew the location of the destination n_d at timestamp 770 and the location was (50, 40).

Current timestamp = 1050

Timestamp is incremented per microsecond basis.

$R_{min} = 5$ km, $R_{max} = 15$ km, $vel_{max} = 1$ km/s while vs = 0.3 km/micros [17].

H = 15.

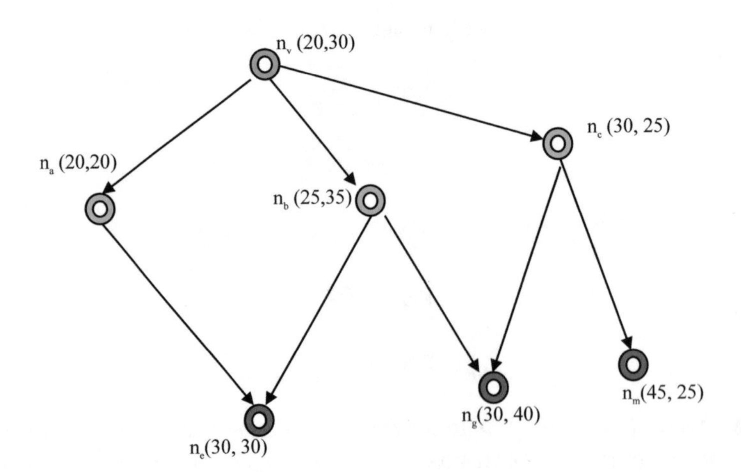

Fig. 6 One example network G rooted at n_v

$$\text{TTL} = (15 \times 20)/(2 \times 0.3) \text{ micros} = 500 \text{ micros.}$$

Assuming RREQ to be initiated at timestamp 620, lifetime already consumed, is (1050 − 620) micros, i.e., 430 micros. Therefore, remaining TTL is (500 − 430) micros, i.e., 70 micros.

Coordinates A, B, Cm and D of LC of n_d are as follows:

A—49.99965, 40.00035
B—50.00035, 40.00035
C—49.99965, 39.99965
D—50.00035, 39.99965

Cartesian distances of all two-hop downlink neighbors of n_v, from A, B, C, and D are shown in kilometers in Table 6.

From Table 6, it is evident that n_e can be completely discarded because it is not possible to reach A, B, C, or D from the node within the stipulated lifetime. So, n_v is supposed to discover least energy route to n_g and n_m only. Let $p_{recv}(a)$, $p_{recv}(b)$, $p_{recv}(c)$, $p_{recv}(e)$, $p_{recv}(g)$ and $p_{recv}(m)$ have values 2, 1, 2, 2, 2 and 2 mj, respectively; $\beta = 3$ and $j = 2$. Energy consumption in the path $n_v \rightarrow n_b \rightarrow n_g$ is given by $(2 \times 50/3 + 1 \times 50/3)$ mj, i.e., 50 mj. On the other hand, energy consumption in the path $n_v \rightarrow n_c \rightarrow n_g$ is $(2 \times 125/3 + 2 \times 225/3)$ mj, i.e., 233.33 mj. Therefore, the path $n_v \rightarrow n_b \rightarrow n_g$ is elected for communication from n_v to n_g. There is only one path to n_m that is, $n_v \rightarrow n_c \rightarrow n_m$ and the corresponding energy cost is 233.33 mj (Fig. 7).

$$\text{OVC}_{G'}(v) = (50 + 233.33)\text{mj} = 283.33\,\text{mj}$$

i.e., $\text{OVC}_G(v) = (200 + 50 + 50 + 233.33 + 233.33)\text{mj} = 766.66\,\text{mj}$

Table 6 Distances of n_e, n_g and n_m from A, B, C, D, of LC of n_d

Two-hop downlink neighbor id	Distance from A, B, C, D of LC of n_d (km)	Required time to reach (μs)
n_e	A—22.359	74.53
	B—22.3607	74.53
	C—22.3528	74.51
	D—22.3607	74.54
n_g	A—19.99965	66.67 (<70)
	B—20.00035	66.67 (<70)
	C—19.99965	66.67 (<70)
	D—20.00035	66.67 (<70)
n_m	A—15.811	52.71 (<70)
	B—15.811	52.71 (<70)
	C—15.811	52.71 (<70)
	D—15.811	52.71 (<70)

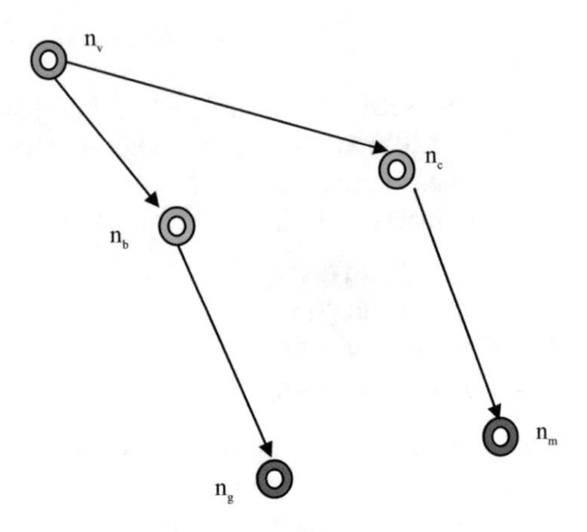

Fig. 7 Optimized network G′

Improvement produced by $G' = \{(766.66-283.33)/766.66\} \times 100\% = 63.04\%$ This humbly states quite an incredible movement.

6 Simulation Results

For simulation, we have used ns-2 simulator in Linux environment. The processor used is 800 MHz Pentium 4 and hard disk capacity is 40 GB. Graphical results are presented in Figs. 8, 9, 10, 11, 12, 13, 14 and 15 to support improvement in favor of E^2BR^2. Please note that E^2BR^2 can be used for route discovery along with any routing algorithm. In this particular paper, we have compared ordinary flooding based route discovery version of the protocols AODV and ABR with E^2BR^2 embedded versions of, i.e., E^2BR^2–AODV, E^2BR^2-ABR, respectively. Number of nodes are taken as 10, 20, 30, 50, 70, 80, 90, 100. Speeds of nodes are taken in the range (0.01, 0.02, 0.04, 0.05, 0.06, 0.07, 0.09 and 0.1 km/s). Transmission range varied between 5 and 15 km. Used traffic type is constant bit rate. Mobility models used in different simulation runs are random waypoint and Gaussian. Simulation metrics are as follows:

Average energy consumption in mj
It is formulated as $\sum_{n_i \in N} \text{eng}(i)/|N|$
where eng(i) is energy consumed by node n_i and N is the set of all nodes in the network.

Average cost of messages
It is formulated as $\sum_{n_i \in N} \text{msg}(i)/|N|$
where msg(i) is number of messages generated and/or forwarded by node n_i and N is the set of all nodes in the network.

Data packet delivery ratio
It is formulated as $\sum_{n_i \in N} \{(|\text{dat-sent}(i)| - |\text{dat-recv}(i)|)/(|\text{dat-sent}(i)|)\}/\{(|N|) \times 100\}$

where dat-sent(i) is the set of data packets transmitted by n_i and dat-recv is the set of data packets generated by n_i that could properly arrive at their respective destinations.

Discovered destinations ratio
It is formulated as $\sum_{n_i \in N} \{(|\text{rdisc-gen}(i)| - |\text{rdisc-found}(i)|)/(|\text{rdisc-gen}(i)|)\}/\{(|N|) \times 100\}$
where rdisc-gen(i) is the set of route discovery sessions generated by n_i and rdisc-found is the number of occasions in which destination was successfully found.

Average delay
It is formulated as $\sum_{n_i \in N} \{\sum_{\lambda \in \text{dat - recv(i)}} ((t_{dest}(\lambda) - t_{start}(\lambda))/(|N| \times |\text{dat} - \text{recv}(i)|)\}$ where $t_{dest}(\lambda)$ and $t_{start}(\lambda)$ denote the timestamps at which a packet λ arrived at destination and the timestamp at which λ began its journey.

Hereby, we humbly state contributions of our present energy-efficient E^2BR^2. To the best of the authors knowledge, it is the first one to focus on the broadcasting of specifically RREQ packets. All state-of-the-art routing algorithms in the literature

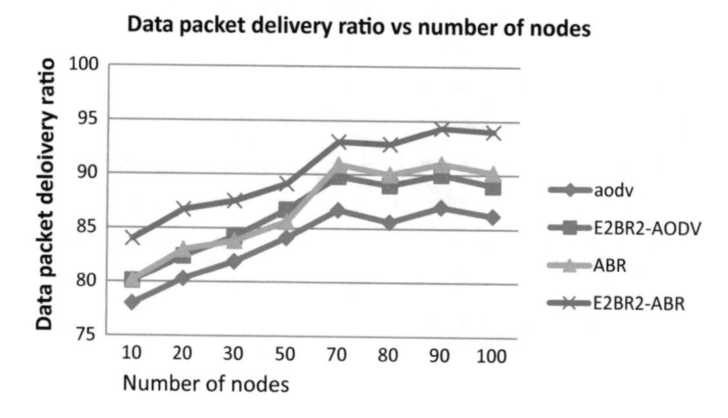

Fig. 8 Data packet delivery ratio versus number of nodes

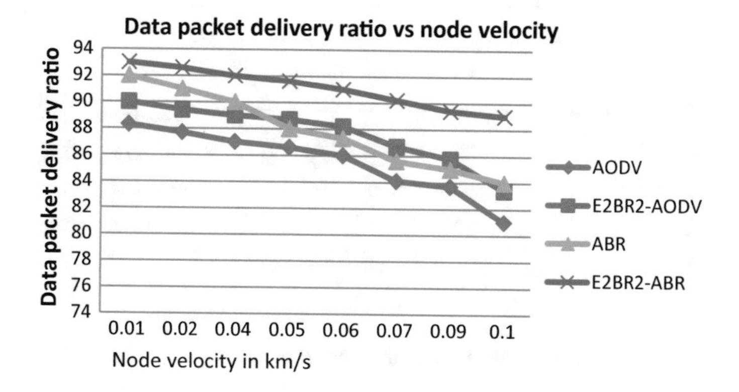

Fig. 9 Data packet delivery ratio versus node velocity

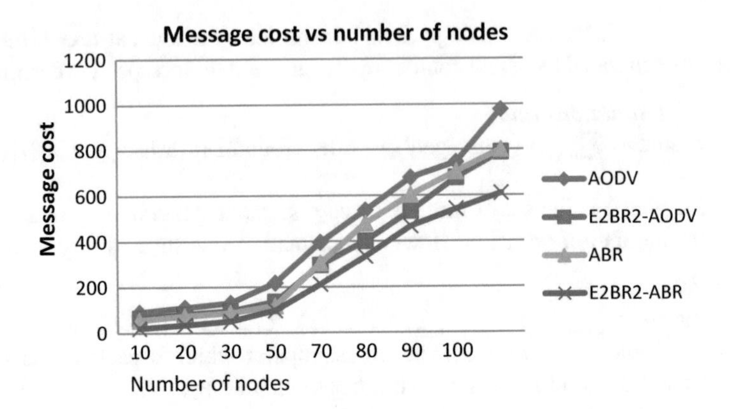

Fig. 10 Message cost versus number of nodes

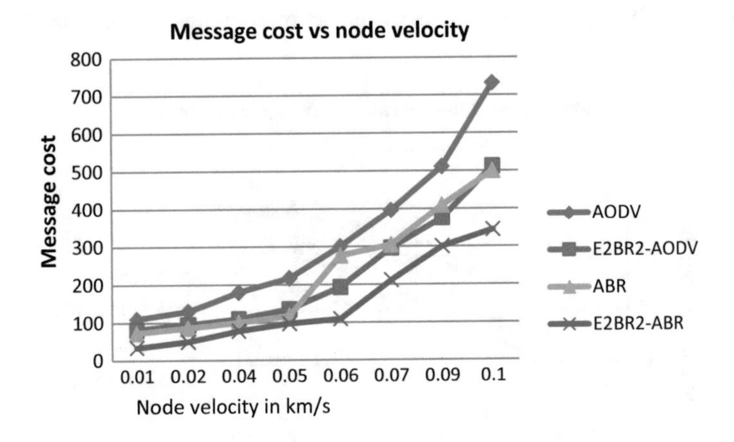

Fig. 11 Message cost versus node velocity

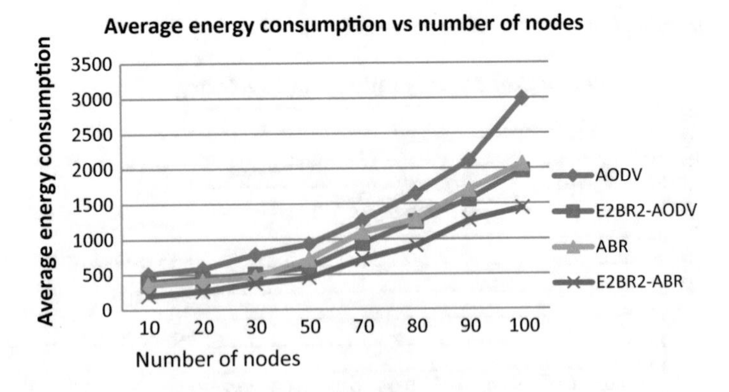

Fig. 12 Average energy consumption versus the number of nodes

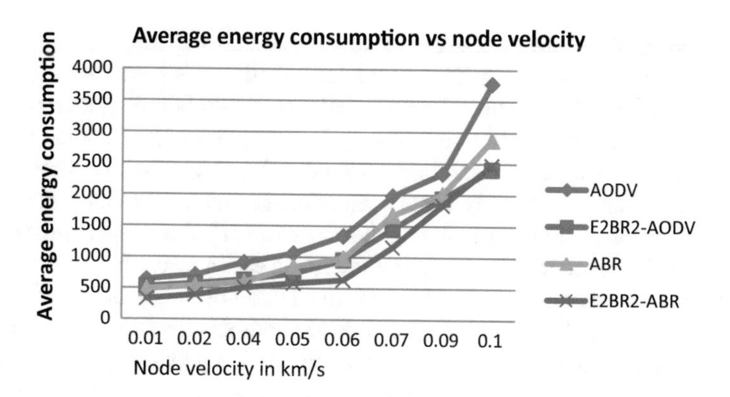

Fig. 13 Average energy consumption versus node velocity

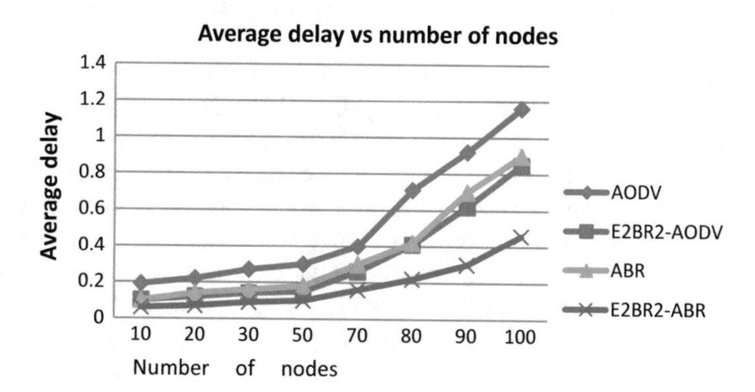

Fig. 14 Average delay versus the number of nodes

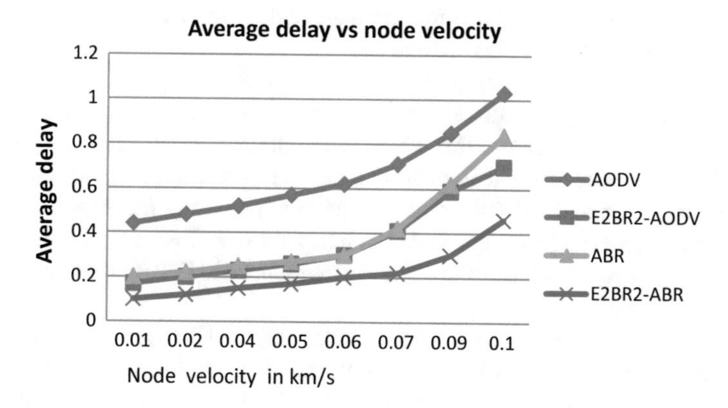

Fig. 15 Average delay versus node velocity

follow flooding for route discovery where each node forwards RREQ packets to each and every neighbor irrespective of the fact whether it is possible to at all reach the destination within a stipulated time. E^2BR^2 uses two-hop neighborhood information to determine the set of two-hop downlink neighbors to which the broadcast message will be forwarded. Examining redundancies in two-hop neighborhood of a node and eliminating them, reduces a significant number of redundancies in the network. Non-destination one-hop neighbors without any downlink connection, are straightaway discarded leading to energy saving. In case a recent location of the destination is not known, energy-efficient paths to all two-hop neighbors are discovered, which preserves a great amount of energy. More optimization is possible if a recent location of the destination is known. Depending upon a previously known location of the destination, the two-hop neighbors from which it is impossible to reach the destination within the lifetime of RREQ packets are discarded and energy-efficient routes are discovered to the remaining set of two-hop neighbors. All these lead to great energy saving.

In E^2BR^2 embedded versions of the protocols, nodes live longer and links break infrequently giving rise to higher throughput, as shown in the Figs. 8 and 9. This, in turn, saves rebroadcasts that would have been otherwise required for repairing broken links, that is, injection of a huge number of RREQ packets unnecessarily consuming energy in nodes. Actually, it is an ominous circle. If you waste energy and leave the network with a very few numbers of alive nodes then it not only generates link breakage but also partitions the network. Resolving that will require flooding RREQ in the network wasting more energy leaving a fewer number of alive nodes, and consequently fewer number of packets successfully delivered to the destination. Therefore, the packet delivery ratio is high in E^2BR^2 embedded protocols and message cost is low as seen in Figs. 8, 9, 10, and 11. E^2BR^2 is equipped with energy-preserving mechanisms described earlier. So, E^2BR^2-AODV and E^2BR^2-ABR produce lesser energy consumption than AODV and ABR, respectively. This is graphically illustrated in Figs. 12 and 13.

However, it is seen for all the protocols that when the number of nodes increases, network connectivity gets better as a result of which, packet delivery ratio increases till the network area saturates after which it definitely starts to get down. As far as node velocity is concerned, it is quite evident that when node velocity increases, new links are made and old links break frequently increasing message cost, energy consumption decreasing average packet delivery ratio. E^2BR^2 embedded protocols save time for link repair. Hence, the delay is also reduced in E^2BR^2 versions as shown in the Figs. 14 and 15.

AODV and ABR apply flooding for broadcasting of RREQ. On the other hand, E^2BR^2-AODV and E^2BR^2-ABR apply techniques of E^2BR^2 for route discovery. Both produce very similar discovered destination ratio for different numbers of nodes and node velocity. Therefore, in spite of so many optimizations in terms of message cost, energy consumption, packet delivery ratio and delay, the number of destinations discovered is no less than flooding, as shown in Figs. 16 and 17.

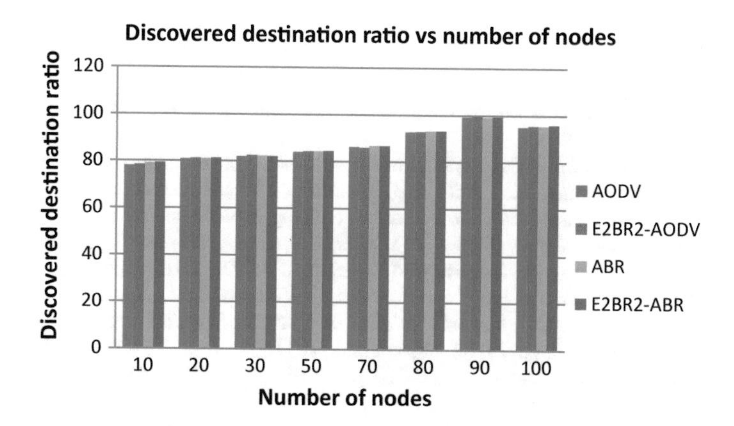

Fig. 16 Discovered destination ratio versus the number of nodes

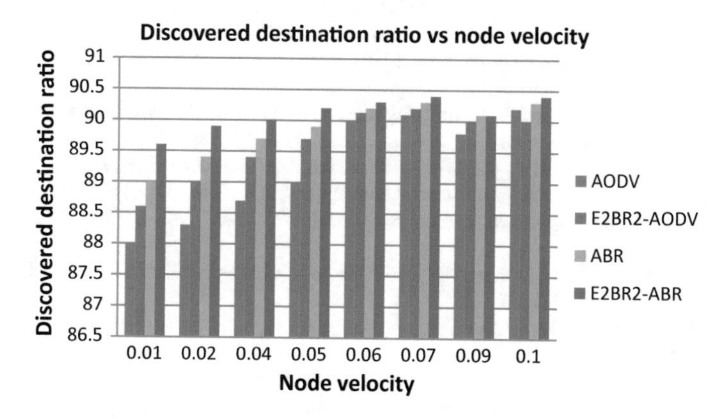

Fig. 17 Discovered destination ratio versus node velocity

7 Conclusion

The present article focuses on an energy-preserving innovative RREQ forwarding mechanism that takes care of redundancies in two-hop neighborhood of nodes, energy-efficient routes to each of the relevant downlink neighbors and discarding certain two hop downlink neighbors based on a practical advantage of knowing a recent location of the destination. The main advantage of the proposed scheme E^2BR^2 is that it can be utilized with any standard unicast routing protocol in ad hoc networks. Instead of applying flooding for route-request, E^2BR^2 should be used. Therefore, the applicability of the proposed scheme is enormous. It is the first one to bring energy efficiency as far as the broadcasting of route-request in ad hoc networks is concerned. It isthe only competitor is flooding which is by default embedded in ordinary version of the routing protocols. In the future, we shall concentrate on investigating the broadcasting of data packets to inculcate energy efficiency as much as possible.

References

1. M. Bala, H. Kaur, Review on routing protocols in mobile ad hoc networks. Int. J. Adv. Res. Comput. Sci. **8**(4) (2017)
2. M.H. Hassan, R.C. Muniyandi, An improved hybrid technique for energy and delay routing in mobile ad-hoc networks. Int. J. Appl. Eng. Res. **12**(1), 134–139 (2017)
3. M. Elhoseny, A.E. Hassanien, Mobile object tracking in wide environments using WSNs, in *Dynamic Wireless Sensor Networks. Studies in Systems, Decision and Control*, vol. 165 (Springer, Cham, 2019), pp. 3–28. https://doi.org/10.1007/978-3-319-92807-4_1
4. M. Elhoseny, A.E. Hassanien, Expand mobile WSN coverage in harsh environments, in *Dynamic Wireless Sensor Networks. Studies in Systems, Decision and Control*, vol. 165 (Springer, Cham, 2019), pp. 29–52. https://doi.org/10.1007/978-3-319-92807-4_2
5. M. Elhoseny, A.E. Hassanien, Hierarchical and clustering WSN models: their requirements for complex applications, in *Dynamic Wireless Sensor Networks. Studies in Systems, Decision and Control*, vol. 165 (Springer, Cham, 2019), pp. 53–71. https://doi.org/10.1007/978-3-319-92807-4_3
6. M. Elhoseny, A.E. Hassanien, Extending homogeneous WSN lifetime in dynamic environments using the clustering model, in *Dynamic Wireless Sensor Networks. Studies in Systems, Decision and Control*, vol. 165 (Springer, Cham, 2019), pp. 73–92. https://doi.org/10.1007/978-3-319-92807-4_4
7. M. Elhoseny, A.E. Hassanien, Optimizing cluster head selection in WSN to prolong its existence, in *Dynamic Wireless Sensor Networks. Studies in Systems, Decision and Control*, vol. 165 (Springer, Cham, 2019), pp. 93–111. https://doi.org/10.1007/978-3-319-92807-4_5
8. A. Banerjee, P. Dutta, Fuzzy-controlled rebroadcasting in mobile ad hoc networks, in *Proceedings of the World Congress on Engineering 2010, WCE 2010*, vol. I, 30 June–2 July 2010, London, UK (2010)
9. Q. Zhang, D.P. Agarwal, Dynamic probabilistic broadcasting in MANETs. J. Parallel Distrib. Comput. **65**, 220–233 (2005)
10. A. Kush, P. Gupta, R. Kumar, Performance comparison of wireless routing protocols. J. CSI **35**(2) (2005)
11. B. Williams, T. Camp, Comparison of broadcasting techniques for mobile ad hoc networks, in *Proceedings of the ACM Symposium on Mobile Ad Hoc Networking and Computing (MOBIHOC)* (2002), pp. 194–205
12. W. Peng, X. Lu, AHBP: an efficient broadcast protocol for mobile ad hoc networks. J. Sci. Technol. (Beijing, China) (2002)
13. T. Clausen, P. Jacquet, A. Laouiti, P. Minet, P. Muhlethaler, A. Qayyum, L. Viennot, Optimized link state routing protocol. Internet Draft: draft-ietf-manet-olsr-06.txt, September 2001
14. J. Sucec, I. Marsic, An efficient distributed network-wide broadcast algorithm for mobile ad hoc networks. CAIP Technical Report 248, Rutgers University, September 2000
15. X. Yuan, M. Elhoseny, H.K. El-Minir, A.M. Riad, A genetic algorithm-based, dynamic clustering method towards improved wsn longevity. J. Netw. Syst. Manag. (Springer US) **25**(1), 21–46 (2017). https://doi.org/10.1007/s10922-016-9379-7
16. M. Elhoseny, X. Yuan, Z. Yu, C. Mao, H. El-Minir, A. Riad, Balancing energy consumption in heterogeneous wireless sensor networks using genetic algorithm. IEEE Commun. Lett. (IEEE) **19**(12), 2194–2197 (2015). https://doi.org/10.1109/LCOMM.2014.2381226
17. E. Royer, C. Toh, A review of current routing protocols for ad-hoc mobile wireless networks, in *IEEE Personal Communications* (1999)
18. W. Peng, X. Lu, On the reduction of broadcast redundancy in mobile ad hoc networks, in *Proceedings of MOBIHOC* (2000)
19. P. Chenna Reddy, P. Chandrasekhar Reddy, Performance analysis of adhoc network routing protocols. Acad. Open Internet J. **17** (2006). ISSN 1311-4360
20. S. Vijay, S.C. Sharma, S. Kumar, Research reviews of IEEE 802.11 wireless ad-hoc networks. Proc. Int. J. Trends Eng. **1**(2), 234 (2009)

21. R. Sivakumar, P. Sinha, V. Bharghavan, CEDAR: core extraction distributed ad hoc routing. IEEE J. Sel. Areas Commun. **17**(8), 1454–1465 (1999)
22. J. Broch, D.A. Maltz, D.B. Johnson, Y.C. Hu, J. Jetcheva, A performance comparison of multi-hop wireless ad-hoc networking routing protocols, in *Proceedings of the 4th International Conference on Mobile Computing and Networking (ACM MOBICOM'98)*, October 1998, pp. 85–97
23. R. Misra, C.R. Manda, Performance comparison of AODV/DSR on-demand routing protocols for ad hoc networks in constrained situation, in *IEEE ICPWC* (2005)
24. M. Elhoseny, A. Farouk, N. Zhou, M.-M. Wang, S. Abdalla, J. Batle, Dynamic multi-hop clustering in a wireless sensor network: performance improvement. Wirel. Pers. Commun. (Springer US) **95**(4), 3733–3753. https://doi.org/10.1007/s11277-017-4023-8
25. M. Elhoseny, X. Yuan, H.K. ElMinir, A.M. Riad, An energy efficient encryption method for secure dynamic WSN. Secur. Commun. Netw. (Wiley) **9**(13), 2024–2031 (2016). https://doi.org/10.1002/sec.1459
26. M. Elhoseny, H. Elminir, A. Riad, X. Yuan, A secure data routing schema for WSN using Elliptic Curve Cryptography and homomorphic encryption. J. King Saud Univ. Comput. Inf. Sci. (Elsevier) **28**(3), 262–275 (2016). http://dx.doi.org/10.1016/j.jksuci.2015.11.001
27. J.C. Bermond, P. Michallon, D. Trystram, Broadcasting in wraparound meshes with parallel monodirectional links. Parallel Comput. **18**(6), 639–648 (1992)
28. M. Mauve, J. Widmer, H. Hartenstein, A survey on position-based routing in mobile ad-hoc networks. IEEE Netw. Mag. **15**(6), 30–39 (2001)
29. T. Pandey, S. Solanki, R. Dubey, Improved performance of AODV routing protocol with increasing number of nodes using traveling salesman problem. Int. J. Comput. Appl. (0975–8887) **98**(16) (2014)

Decision Support System for Smart Grid Using Demand Forecasting Models

Sonali N. Kulkarni and Prashant Shingare

Abstract In recent years, the penetration of renewable energy (RE) into the power system is ever increasing to meet the exponential increase in power demand. The number of prosumers, generating renewable energy in a distributed manner and participating in a power network is also increasing drastically. This has posed serious issues to grid stability, as instantaneous power demand and renewable power generation are inherently intermittent and dynamic in nature. Precise demand–supply balance is critical but essential for maintaining the stability of the grid. In order to accommodate excess penetration of RE and maintain demand–supply balance, a detailed revision in infrastructure and planning or smart grid implementation becomes essential. In this work, we have designed an innovative hybrid ARMA demand forecast model, using historical power demand of Maharashtra state in India. The forecast results of hybrid ARMA model are compared with traditional statistical models. A precise power demand balance is key to smooth, stable and reliable operation of smart grid or modern power network, models designed in this work shall be useful to smart grid: energy management system in decision-making processes related to real-time operation and control of power system.

Keywords Decision support system · Demand–supply balance · Forecasting for smart applications · Power quality · Smart grid · Time series analysis

S. N. Kulkarni (✉)
Electronics & Telecom Engineering, Rajiv Gandhi Institute of Technology, Versova, Andheri (W), Mumbai 400053, Maharashtra, India
e-mail: ksonali2006@gmail.com

P. Shingare
Vertiv Energy Pvt. Ltd, NITCO Business Park, Wagle Industrial Estate, Thane (W) 400604, Maharashtra, India
e-mail: prashant.shingare@vertivco.com

M. Elhoseny and A. K. Singh (eds.), *Smart Network Inspired Paradigm and Approaches in IoT Applications*, https://doi.org/10.1007/978-981-13-8614-5_4

1 Introduction

A typical modern-day power infrastructure consists of the interconnection of different types of power generators both conventional and renewable, transmission lines, transformer, and varying nature of load. The power system structure and operations become more complex due to the integration of different types of loads and power generators [1]. The renewable energy (RE) sources like wind and solar offers alternative sources of energy and becoming more popular as they are pollution free, technologically effective and provides electricity without giving rise to carbon dioxide emissions. But intermittent and unpredictable renewable energy generation coupled with varying nature of load causes demand–supply mismatch and impacts grid stability [1–5]. In India, the existing electricity grid is not capable enough to handle excess penetration of RE. Also, existing electricity grid has no potential to offer adequate services addressing energy efficiency, reliability, and security and the integration of RE at the scale needed to meet clean energy demand for future [1, 3]. Therefore, in order to handle the penetration of RE energy, implementation of smart grid technology is essential. The smart grid concept is still in the developing stage, though it is having many challenges. The pilot deployments are going on at some places globally. Also, the introduction of smart grid reduces overall greenhouse gas (GHG) emissions with demand management that encourages energy efficiency, improves reliability, and manages power more efficiently and effectively. A smart grid is also capable to address growing penetration of renewable energy and achieve a balance between demand and generation. In smart grid scenario, short-term power demand forecasting techniques discussed in this paper shall be useful as a decision support system while making decisions related to power purchase, load switching, generation planning, and scheduling to maintain balance between demand and generation [1, 5]. The power system operators rely on demand forecasts results and take decision to plan and achieve optimum generation to maintain healthy power quality, reliability in an economic sense [5].

Presently, each state load dispatch centers across India (SLDC) schedule their power demand to regional load dispatch centers (RLDCs), mainly based on their experience or heuristic knowledge. In this paper, we have discussed smart grid architecture and short-term power demand forecasting models that are useful as a reference for various decision-making steps in a smart grid. The hourly power demand data available at Maharashtra SLDC is used to design hybrid ARMA and different traditional statistical models to generate and compare short-term demand forecasts with the help of various performance parameters. The forecast results shall be useful to load dispatch centers for projecting power demand to the power system operator. In smart grid applications, forecast models shall provide backend support to the energy management system to achieve smooth and stable grid operation.

This paper is organized as follows: Sect. 1 reviews the architecture and functions of smart grid components and challenges in smart grid implementation. Section 2 explains demand forecasting types, difficulties, challenges, and its importance in power system. Section 3 explains different short-term demand forecasting techniques, their mathematical models and results followed by the conclusion.

1.1 Smart Grid Components

Smart grid is a modern electric power grid, that enhances efficiency and reliability of power system through automatic control, highly efficient power converters, latest communication facilities, modern sensing and metering technologies, and energy management techniques based on optimization of demand, generation, energy and network usability, and so on [6–8]. With the incorporation of automatic control and modern communications methods [9], smart grid architecture facilitates improved flexibility, accessibility, reliability, efficiency, and robustness to the power system. The key factors involved in the operation and planning of a smart grid power system are distributed network operators (DNOs), distributed generators (DGs) and a variety of electricity consumers. The efficient and optimized operation of the smart grid is possible due to the energy management system (EMS) that uses modern control technologies to achieve dynamic control of various elements present in the smart grid network. The versatile architecture of smart grid and highly interacting information flow among smart grid components are shown in Fig. 1 [7, 9].

1. Supervisory Control and Data Acquisition (SCADA) system: The SCADA receives measurement data from remote data collecting devices (RTUs) placed at different conventional and RE generators located at remote locations, as a part of smart grid network and transmits it to EMS [6, 9]. SCADA systems monitor various parameters related to equipment and its states like power production, inverter output, inverter status, AC grid conditions, weather station data, temperature of key components, solar irradiance, etc. In certain applications, SCADA coordinates wind turbine outputs and provides reactive power support for utility system whereas in some wind farms; SCADA-authorized users can modify different parameters of wind energy converter (WEC) controllers voltage control system (VCS), etc. Thus, SCADA system is of special importance in smart grid and wind farm controllers because it helps to maintain optimized performances under varying power and load conditions [10].
2. Remote Terminal Units (RTUs): In smart grid, remote terminal unit is used to monitor and control different power generation units and distributions remotely through advanced sensors. RTU provides electricity parameter information to SCADA using a combination of different modern communication techniques or Internet of Things (IoT).
3. Energy Management System (EMS): EMS is a heart of smart grid as it monitors and controls functions of smart grid components [10]. EMS uses SCADA data, state estimation algorithm (SEA), historical data, generation, and load forecast

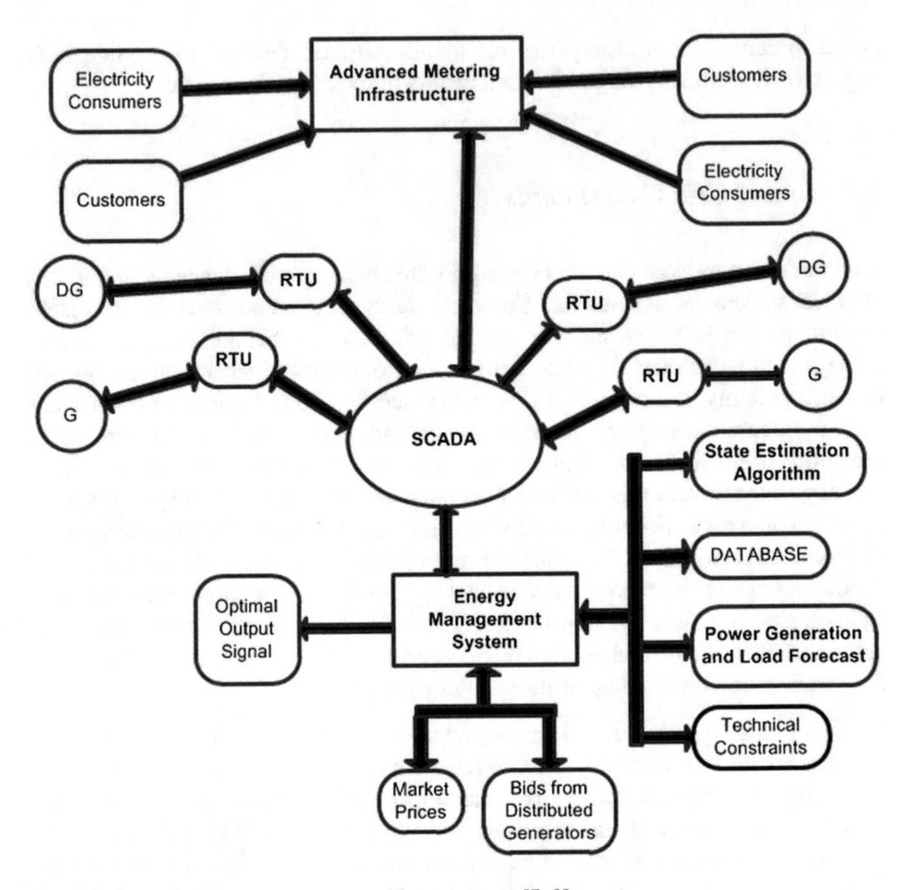

Fig. 1 Information flow among smart grid components [7, 9]

(GLFS) to analyze and decide necessary actions required to achieve the optimal combination of control variables for active and reactive power control by changing position of On Load/No Load Tap Changer (OLTC) transformer in a substation.

4. Advanced Metering Infrastructure (AMI): The advanced metering infrastructure consists of smart meters with wireless (IoT), wired communication, and/or a combination of both, having bidirectional data flow capabilities [7, 9]. It consists of communications hardware, software, associated system, and data management software that acquire two-way network communications between advanced meters and utility business systems. It results in the collection and distribution of information to customers, retail power supplier and to the utility itself. In some applications, AMI provides customers with real-time (or near real time) pricing of electricity, which supports utilities in load management when necessary [11].

5. State Estimation Algorithms (SEA): SEA is a set of methods and mathematical procedures used to evaluate the state of smart grid network by using remotely measured data. SEA collects real-time information about the grid and received data is used by SEA to identify the current and voltage profiles at all its nodes to achieve full control of grid. Moreover, in order to obtain an accurate state estimation of voltage and current profiles, adequate monitoring systems along with AMI are used [9].
6. Power Generation and Load Forecast System (GLFS): The electric utilities try to balance power generation and demand in order to offer a good service at a competitive price. For this purpose, electric utilities need accurate electric load records [12]. The load and weather forecast information is used for generating forecast data which are used to achieve optimized grid performance [13]. However, electric load depends on many factors like day of the week, the month of the year, etc., which makes load forecasting quite a complex process requiring something other than statistical methods [12]. The reliable and stable of the power grid is achieved by using prior information about load pattern or load curve, data received by SCADA system related to different power plants or generators, power optimization, load distribution, redistribution of power flow between power plants [9]. Energy management system uses SCADA data, demand–generation forecast, considers available technical and infrastructural limitations to take important decisions for real-time operation and control of power network or smart grid. In certain microgrid applications, electric load forecast model is designed using an Artificial Neural Network (ANN), which has the ability to perform Short-Term Load Forecasting (STLF) [12].

1.2 Challenges in Smart Grid Implementation

As per reports on a survey carried out and discussed in [14], challenges to smart grid implementation are categorized on the basis of

1. General conditions and framework
2. Terms for taxes, financial inducements, and incentives, Involvement and role of different stakeholders in the group
3. Know how of market participants

Among all the above categories, the majority challenges belong to general conditions and framework primarily related to standardization, regulation elements, and implementation mechanisms. They are

a. Lack of standards on smart grid technology: Worldwide there are lack of standards, definitions, and well-defined minimum functionalities for smart grid solutions [5, 14],
b. Transmission-line capacity: In most of the cases, available grid networks are not

suitable for integration of RE, so it is required to strengthen grid with sufficient transmission line capacity to connect RE power [14].

c. Network limitations: Difficulty for network operators to introduce more advanced structures in their network in order to use network efficiently [14].
d. Role of stakeholders: There is a need for improvement in the definition and assignment of roles and responsibilities of stakeholders [15].
e. Data security and privacy issues: Security of shared information, among various domains, is essential at different levels in smart grid architecture [14].
f. Developing decentralized architectures: Enabling smaller scale electricity supply systems to operate in coordination with the total system [15].
g. Active demand side: Enabling all consumers, with or without their own generation, to play an active role in the operation of the system [15].
h. Integrating intermittent generation: Need for designing the best ways for integrating intermittent power generation including micro-generation [5, 15].
i. Artificial Intelligence: Maintaining a balance between demand and generation intelligently without affecting power quality [15].
j. Electric vehicles: The major challenge for the smart grid is to accommodate the consumer demand for electric vehicles due to their mobile and highly dispersed character [16].

2 Power Demand Forecasting

Power Demand Forecasting is a process to forecast electricity requirement for different types of power consumers by analyzing historical demand data. The power requirement of different types of power customers such as residential, commercial, and industrial varies from time to time. Therefore, historical information of hourly, daily, weekly, monthly, and yearly electricity demand is analyzed to determine demand forecast [17]. The process of power demand forecasting is being used to forecast demand from decades. In smart grid applications, accurate power demand forecasting information is needed due to the following reasons:

- Accurate demand forecasting helps power system operator in planning operation of electric power systems smoothly. Demand and generation balance is necessary to maintain required grid frequency and hence power quality [18].
- It supports power system operator in making important decisions related to power purchase, fuel allocation, generation scheduling, loads switching, etc. [19].
- Demand forecasting is an important factor that decides the economical operation of power companies [20].
- Accurate power demand forecasting helps in operational and maintenance cost savings, as forecasting error leads to the increased operating cost of the network [19, 21].

Table 1 Types of demand forecasting

Sr. No.	Name	Duration	Used for:
1.	Very short term	Few minutes to an hour	Real-time control
2.	Short term	Few hour to few weeks	Scheduling of power purchase, generation, maintenance
3.	Medium term	Few months to five years prior to the actual requirement	Utility planning, maintenance scheduling and for coordination of power sharing
4.	Long term	Twenty years in advance	Decision on regulatory policies, power prices, capacity planning

- Short-term load forecasting helps to plan, estimate, and manage the load to prevent network overloading. This results into increase in network reliability, reduction in possible occurrences of equipment failures and blackouts [18, 19].
- Thus, demand forecasting plays an important role in the process of decision-making and control in smart grid by considering network bottlenecks or technical limitations if any [8].

2.1 *Types of Power Demand Forecasting*

On the basis of duration, power demand forecasting is broadly classified into four types given in Table 1 [8, 19, 22, 23].

Table 1, also provides applications of different demand forecasts in power system management. Among different methods, short-term load forecasting is of very much importance for smooth, reliable and real-time operational control of utilities in the power network/smart grid. In Maharashtra state of India, SLDC's need to provide short-term power demand forecast with time lead of two–three hour to the central power network operator for real-time control.

2.2 *Challenges in Power Demand Forecasting*

For smooth and reliable operation of the power network, it is very import but crucial to maintaining a balance between demand and generation. As discussed above accuracy of power demand forecast contributes major share in the economical, efficient and reliable operation of smart grid. However, the exact estimation of power demand is extremely complicated due to the involvement of various uncertainties. In the

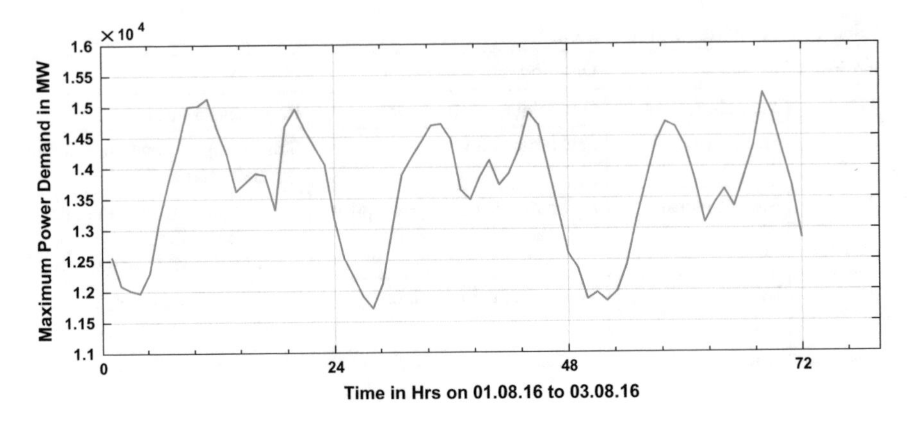

Fig. 2 Hourly power demand (in MW) [24]

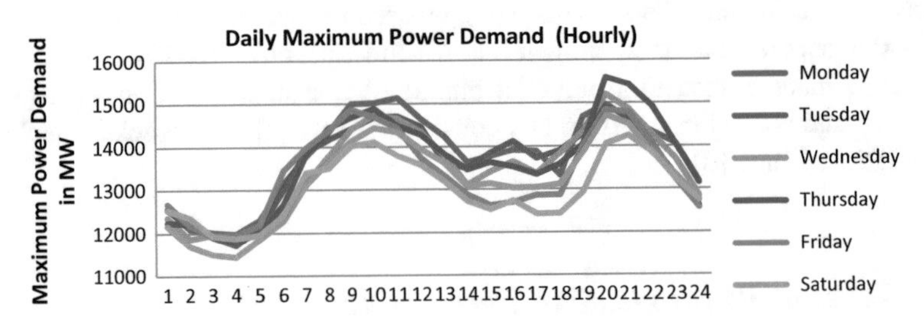

Fig. 3 Hourly power demand (MW) on different days in a week [24]

most competitive electricity market, hourly power demand series has the following characteristics [25, 26]:

- Hourly power demand of different types of electricity consumers varies from time to time, that result in fluctuations in total power demand. The hourly peak power demand of three days, at SLDC Maharashtra, India is shown in Fig. 2.
- The changes in weather conditions like temperature and humidity increases the complexity in demand–supply balance [27].
- Multiple seasonality: The total power demand (in MW) varies with the month in a year, season, week in a month, day of a week, and special occasion or festivals, etc. [20]. Figure 3 shows the variation in total peak power demand observed at SLDC, Maharashtra, India during a particular week.

From Figs. 2 and 3, it is observed that there is high variation in power demand and the frequency of variation is also high. Further power generation is also fluctuating and changes in weather conditions like temperature, humidity and season increases the complexity in the generation and hence demand–supply balance becomes challenging [27].

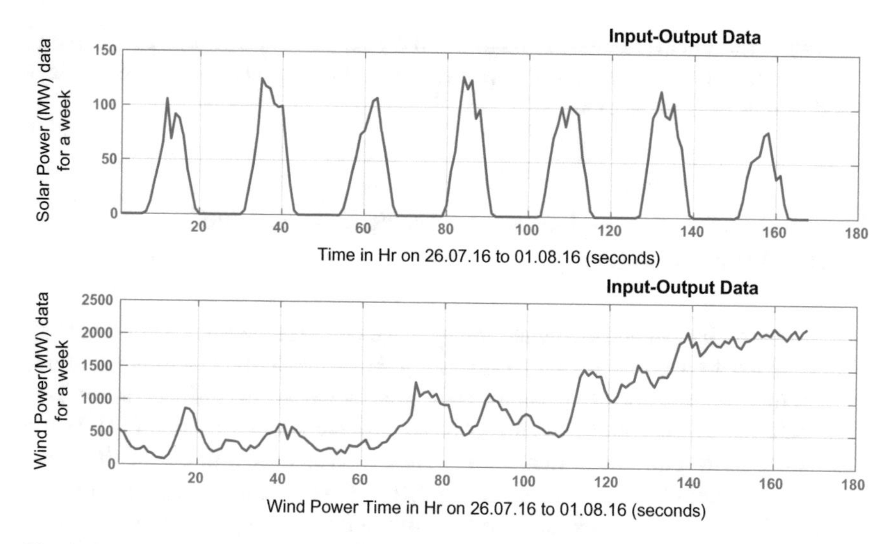

Fig. 4 Hourly solar and wind power generation (MW) per week [24]

The following difficulties are faced at the power generator side, during demand—supply balance.

- The power generated by conventional generators varies with the availability of resources.
- Power generated by most popular renewable energy sources like wind and solar is intermittent and varies from time to time due to intermittent variation in natural resources [1, 2, 18, 28]. Figure 4 shows the hourly variation in solar and wind power generation for the duration of a week.

2.3 *Difficulties in Demand Forecasting*

Power demand continuously varies with time. Therefore, it is very import but crucial to forecast power demand for the smooth functioning of the smart grid. However, the exact estimation of power demand is extremely complicated due to the involvement of various uncertainties. In the most competitive electricity market, hourly power demand series presents the following characteristics [25, 26]:

- Variation in demand and frequency of variation is high.
- Nonschedulable nature of demand.
- Nonconstant mean and variance (nonstationary series).

- Multiple seasonality: power demand (in MW) varies with month, week, and day of a week. It is also observed that power demand during weekends and holidays is different than weekdays.

3 Short-Term Power Demand Forecasting Methods and Model Results

A variety of statistical and artificial intelligence techniques based on different techniques are used for power demand forecasting. For short-term power demand forecasting, different methods like Time Series Analysis, Autoregression, Similar Day Approach, Artificial Neural Network (ANN), Expert Systems are used [19]. While forecasting short-term power demand a variety of factors like time, weather data, possible customer class, etc., are considered. The time factors consider the time of the year, the day of the week, and the hour of the day as the power demand varies with the above factor. Time series analysis is a statistical approach of forecasting that has attracted many scientists and researchers over the past few decades. Time series forecasting methods are used in numerous fields like business, finance, science, and engineering to determine future values [29, 30]. The main aim of time series modeling is to study the past available time series data to develop the appropriate model or time series forecasting is an act of predicting future values by understanding the past [20, 31]. A statistical approach for short-term power demand forecasting uses a mathematical model to represent power demand as a function of different factors such as time, weather, and customer class, etc. This model is then used to generate a power demand forecast. A lot of efforts and care is required to be taken while developing and fitting an adequate model from historical time series data to improve forecasting accuracy. In this paper, we have used a statistical approach to design and generate short-term demand forecasts using historical power demand of Maharashtra state in India. The hourly demand data is selected, preprocessed and converted into the desired format using NumXl. This time series data is used to design short-term demand forecast model to generate a demand forecast.

The most commonly used statistical mathematical models are of two types: additive model and multiplicative model [20, 23, 32, 33]. The additive model designed for power demand forecast is the function of four components and is given in (1),

$$\mathrm{L} = \mathrm{Ln} + \mathrm{Lw} + \mathrm{Ls} + \mathrm{Lr} \tag{1}$$

where L represents total power demand, Ln represents "normal" part of demand, it is a set of standardized load shapes for each "type" of day that has been identified as occurring throughout the year, Lw is weather dependent part of power demand, Ls is a special event parameter which creates a deviation in demand from usual pattern, and Lr is a completely random parameter, called as noise [20, 32, 33].

The multiplicative model is given by (2)

Table 2 Performance of Naïve forecast

Samples	MFE	MAFE	MSE	MPE (%)	MAPE (%)
24	20.78	447.74	250798.08	0.09	3.29
36	6.38	427.66	237832.38	−0.01	3.16
72	3.77	426.93	252470.78	0.03	3.13
192	−8.84	402.23	14918.03	−0.13	2.81

$$L = Ln * Fw * Fs * Fr \tag{2}$$

where Ln is normal or baseload/demand and Fw, Fs, and Fr are correction factors that vary overall power demand and are positive numbers based on weather (Fw), special events (Fs), random fluctuation (Fr). Sometimes, factors like electricity pricing (Fp) load growth (Fg) are also taken into consideration [20, 32]. To determine accurate power demand forecast, it required to study and identify the impact of above factors on demand forecast by decomposing historical time series [17, 20].

3.1 Naive Forecast Method (Same Day/Similar Day Approach)

Naive forecasting method called as classical, is one of the simple and widely used methods in short-term forecasting studies. According to Naive forecast method, the most recently available data is used as a forecast data given by (3) [22, 34].

$$Y_t = Y_{t-1} \tag{3}$$

where Y_t is power demand at time t, Y_{t-1} is the most recent time series historical power demand data. The Naive forecast model is implemented using NumXl and for the various historical samples and the model performance is analyzed on the basis of various performance parameters like MFE, MAFE, MSE, etc., as given in Table 2.

3.2 Exponential Smoothing Method

Exponential smoothing method uses a weighted average of historical time series to calculate the forecast. It is a special case of the weighted moving averages method in which we select only one weight for the most recent observation and the weights for other data values go on decreasing and are calculated automatically as the observations move further into the past. The exponential smoothing equation is given by (4) [33, 34]:

Table 3 Exponential smoothing forecast result comparison

Samples	α	MSE	MPE (%)	MAPE (%)	U-stat
24	1.60	164820.84	+0.17	2.83	0.81
48	1.40	208385.70	−0.15	2.90	0.93
72	1.39	652133.75	−0.13	2.90	0.11

$$F_{t+1} = \propto * Y_t + (1 - \propto) * F_t \quad (4)$$

where F_{t+1} is forecast for the time series for a period of t + 1, Y_t = Actual value of time series in period t, F_t is Forecast of time series for period t, α = Smoothing constant whose value is $0 <= \alpha <= 1$. The value of α is low, when there are less fluctuations demand over the selected period of time and when $\alpha = 1$, each forecasted value is same as previous value, but this does not assure minimum MSE. For historical power demand series, MSE is minimum for value of α can be other than 1. In this paper, we have used NumXl solver to calculate the value of α for minimum value of MSE and this value of α will provide best results as compared to other. Table 3 shows the comparison of exponential smoothing model results for different sample sizes.

3.3 Autoregressive Moving Average (ARMA) Method

Autoregressive Moving Average (ARMA (p q)) on is the most widely used statistical techniques for power demand forecasting. The expression for output Y(t) is given in (5) [33].

$$Y(t) = \sum_{k=1}^{p} Øk * y(t-k) + a(t) + \sum_{j=1}^{q} bj * e(t-j) \quad (5)$$

First term corresponds to autoregressive process of order p and Øk is known as the coefficient of delay polynomial. In autoregressive process AR(p), the current value of time series Y(t) is expressed linearly in terms of its previous values like [Y(t−1), Y(t−2) … Y(t−p)] and the coefficient of delay polynomial Øk. The second term is a random noise a(t) and third term correspond to moving average MA(q) process of order q. The ARMA (p q) model is designed using software tools like NumXl and MATLAB using historical hourly power demand of Maharashtra state in India. The comparison of model forecast results of ARMA (1 17) and ARMA(2 17) are given in Fig. 5. The forecast results of two ARMA models are compared based on different parameters as shown in Table 4. From Table 4, it is observed that the performance of ARMA (2 17) is better as compared to ARMA(1 17).

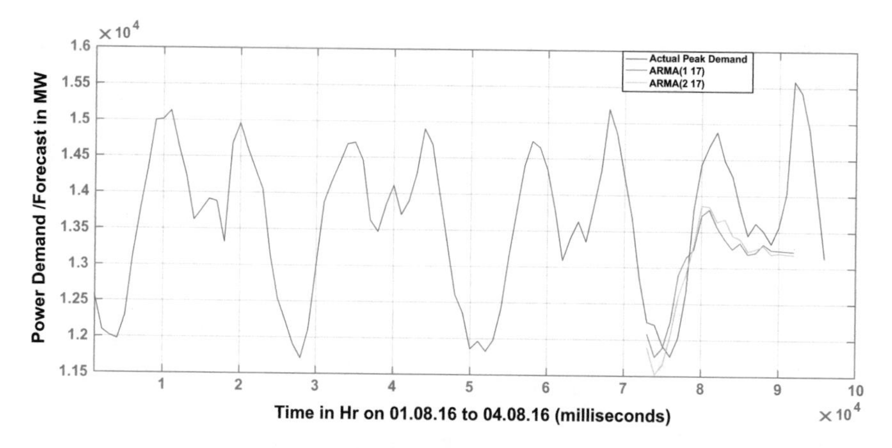

Fig. 5 Demand forecast results of different ARMA models

Table 4 Comparison of ARMA and ARMA_H model results

Model/parameters	Fit to estimation (%)	Forecast prediction error	Mean square error
ARMA_H(7 1)	82.39	124200	28670
ARMA_H(5 1)	78.56	114300	42460
ARMA_H(5 2)	79.92	111800	37270
ARMA_H(6 2)	81.51	120100	31600
ARMA (1 17)	72.17	216300	74800
ARMA (2 17)	73.90	197400	65810

3.4 Hybrid ARMA (p q) Model

In hybrid ARMA model, the time series data is processed and used to design the model for demand forecasting. In hybrid ARMA process, Y'(tn) time series as input to the integrated autoregressive moving average process, Y(tn)d on any date "d" is determined using (6) as

$$\mathrm{Y}'(\mathrm{tn}) = \left[\mathrm{y}(\mathrm{tn})\mathrm{d} - 1 + \mathrm{y}(\mathrm{tn})\mathrm{d} - 2 + \mathrm{y}(\mathrm{tn})\mathrm{d} - 3\right]/3 \quad (6)$$

where n = 1, 2, 3, 4, … 0.24 and d = 1, 2, 3 … 0.31. The output Y(t) of hybrid ARMA process is given in (7)

$$\mathrm{Y}(\mathrm{t}) = \sum_{\mathrm{k}=1}^{\mathrm{p}} \emptyset\mathrm{k} * \mathrm{y}'^{(\mathrm{tn}-\mathrm{k})} + \mathrm{a}(\mathrm{t}) + \sum_{\mathrm{j}=1}^{\mathrm{q}} \mathrm{bj} * \mathrm{e}'(\mathrm{tn} - \mathrm{j}) \quad (7)$$

where

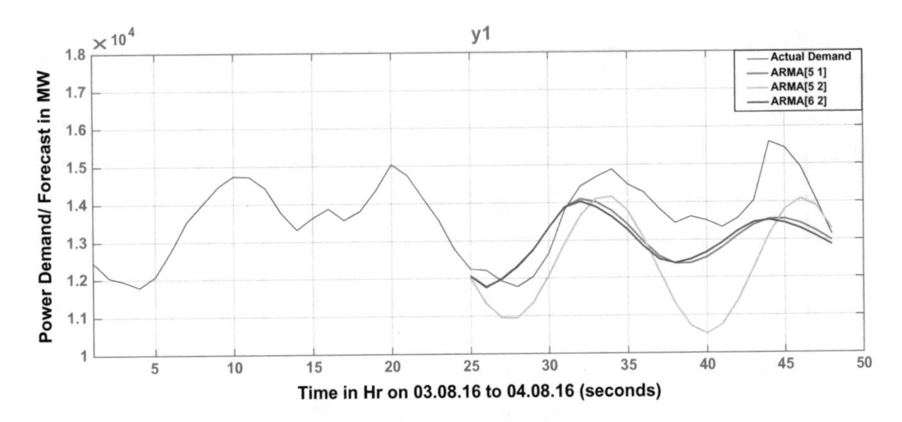

Fig. 6 Hybrid ARMA model demand forecast result comparison

$$\begin{aligned} y'(tn-1) &= \left[y(tn-1)d-1+y(tn-1)d-2+y(tn-1)d-3\right]/3 \\ y'(tn-2) &= \left[y(tn-2)d-1+y(tn-2)d-2+y(tn-2)d-3\right]/3 \\ &\vdots \\ y'(tn-p) &= \left[y(tn-p)d-1+y(tn-p)d-2+y(tn-p)d-3\right]/3 \end{aligned}$$

The hybrid ARMA (p q) model, data preprocessing is carried out using software like NumXl and MATLAB is used for the model design. By observing ACF and PACF plots of power demand data, model parameters are obtained. The forecast output of hybrid ARMA_(5 1), ARMA_H(5 2), ARMA_H(7 1), and ARMA_H(6 2) models are compared on the basis of various performance parameters like MES, fit to estimation and FPE as shown in Table 4. The forecast model output of ARMA_H(7 1) and ARMA_H(6 2) are shown in Fig. 6.

4 Conclusion

Smart grid pilot system deployment is on the rise globally due to its ability to accommodate excess penetration of distributed renewable energy along with conventional generators. In this paper, we discussed the challenges and difficulties in smart grid implementation. Further, we discussed smart grid architecture; the importance of precise demand—supply balance to maintain grid stability and power quality; and use of demand forecasting models. Also, we discussed the significance of demand forecast models in energy management system while taking important decision to maintain demand supply balance for smooth and reliable operation of the smart grid. Next, we discussed and presented short-term power demand forecasting models using conventional methods like Naïve, Exponential smoothing and ARMA(p q) and their performances are compared on the basis of different performance parameters. All models are designed using an exactly identical historical power demand data obtained for the state of Maharashtra in India. From Naïve forecast methods results, it is observed that the performance of this method for a higher number of historical data points is better due smaller value of MSE. Also, the performance of exponential smoothing method is better for less number of historical data points. Further, we discussed illustrative numerical examples by designing two ARMA models based on the actual data obtained from the Maharashtra SLDC centre India. From the simulation results of ARMA models, it is observed that the performance of ARMA(2 17) is better than ARMA(1 17). Further, in this paper, we proposed a novel method for time series data analysis named hybrid ARMA model to forecast power demand. The demand forecast results of four models ARMA_H(7 1), ARMA_H(5 1), ARMA_H(5 2) and ARMA_H(6 2) models are compared with ARMA models on the basis of performance parameters. It is observed that the performance of hybrid ARMA_H(6 2), ARMA_H(5 2) models are almost similar. The performance of ARMA_H (6 2) model is better as compared to all other models in designed time span. The forecast models designed in this work shall support smart grid: EMS during decision-making processes for real-time operation and control of utilities; to maintain power quality, grid stability, and power economics.

Acknowledgements Professor Sonali N. Kulkarni is thankful to her colleagues, Principal and Management of Bharati Vidyapeeth College of Engineering, Navi Mumbai, Maharashtra, India for supporting and encouraging her during this research work.

References

1. Z. Lubosny, *Wind Turbine Operation in Electric Power Systems* (Springer, Berlin, 2003), pp. 11–30. ISBN 3-540-40340-X
2. M. Sandhu, T. Thakur, Issues, challenges, causes, impacts and utilization of renewable energy sources—grid integration. Int. J. Eng. Res. Appl. **13**(3), 636–643 (2014), www.ijera.com. Version 1. ISSN: 2248-9622
3. T. Ackermann, A book on wind power in power systems, in *Wind Power* (Wiley, England, 2005), pp. 97–112, 169–182 (6, 9)
4. P. Shingare, Recent technology trend in wind energy grid integration, in *Invited Technical Talk at One Week Workshop on, "Advances in Power System"*, VJTI, Mumbai, 16–21 May 2016
5. S.N. Kulkarni, P. Shingare, A review on power quality challenges in renewable energy grid integration. Int. J. Curr. Eng. Technol. E-ISSN 2277–4106, P-ISSN 2347–5161
6. V. Gungor, G. Hancke, C. Buccella, Smart grid technologies: communication technologies and standards. IEEE Trans. Ind. Inform. **7**(4) (2011)
7. Focus Group on "Smart Grid overview", International Telecommunication Union Telecommunication Standardization Sector Study Period 2009–2012, Dec 2011
8. S.N. Kulkarni, P. Shingare, A review on smart grid architecture and implementation challenges, in *International Conference on Electrical, Electronics, and Optimization Techniques (ICEEOT)* (2016), pp. 1–6 (978-1-4673-9939-5/16 ©2016)
9. S. Pierluigi, P. Antonio, R. Gerasimos, *Wind Turbines Allocation in Smart Grids* (IEEE, 2013) (978-1-4799-0224-8/13/2013)
10. F. Richard Yu, P. Zhang, Communication systems for grid integration of renewable energy resources. Netw. IEEE (IEEE Communication Society) **25**(5) (2011)

11. NIST, Framework and Roadmap for Smart Grid Inter-operability Standards, Release 1.0, US Department of Commerce, Gary Locke, Secretary
12. L. Hernandez, C. Baladrn, Short-term load forecasting for microgrids based on artificial neural networks. Energies (2013)
13. F.T. Aula, S.C. Lee, Grid power optimization based on adapting load forecasting and weather forecasting for system which involves wind power systems. Smart Grid Renew. Energy (Scientific Research) (2012)
14. L. Hernandez, C. Baladrn, *CEER Status Review of Regulatory Approaches to Smart Electricity Grids* (Council of European Energy Regulators ASBL, 2013)
15. L. Hernandez, C. Baladrn, *Smart Grids: Strategic Deployment Document for Europe's Electricity Network of the Future* (European Technology Platform, 2008). http://www.smartgrid.eu/documents/smart
16. T. Vijayapriya, D.P. Kothari, Smart grid: an overview. Smart Grid Renew. Energy (Scientific Research) (2011)
17. E. Almeshaiei, H. Soltan, A methodology for electric power load forecasting. Alexandria Eng. J. **50**, 137–144 (2011)
18. A. Velayutham, Expert talk on Power Quality (PQ) issues in smart grid and renewable energy soures. Ex Member, MERC, at SGRES, CPRI, Bangalore (2015)
19. E.A. Feinberg, D. Genethliou, Load forecasting, in *Applied Mathematics for Restructured Power Systems* (Springer, Berlin, 2005), pp. 269–285
20. G.E.P. Box, G.M. Jenkins, G.C. Reinsel, *Time Series Analysis—Forecasting and Control*, vol. 33, 4th edn. (Wiley, New York, 1982), pp. 533–545
21. T. Haida, S. Muto, Regression based peak load forecasting using a transformation technique. IEEE Trans. Power Syst. **9**, 1788–1794 (1994)
22. C. Nataraja, M.B. Gorawar, G.N. Shilpa, J. Shri Harsha, Short term load forecasting using time series analysis: a case study for Karnataka, India. Int. J. Eng. Sci. Innov. Technol. (IJESIT) **1**(2), 1–9 (2012)
23. D.W. Bunn, E.D. Farmer, *Review of Short-term Forecasting Methods in the Electric Power Industry*, vol. 33 (Wiley, New York, 1982), pp. 533–545
24. maha_state_sldc.in_dailyreport
25. M. Negnevitsky, P. Mandal, A.K. Srivastava, *Machine Learning Applications for Load, Price and Wind Power Prediction in Power Systems* (IEEE, 2009)
26. N. Amjady, M. Hemmati, Energy price forecasting. IEEE Power Energy Mag. 20–29 (2006)
27. K. Darrow, B. Hedman, The role of distributed generation in power quality and reliability. Final Report for New York State Energy Research and Development Authority, 17 Columbia Circle, Albany, New York 12203–6399 (2005)
28. M. Singh, V. Khadkikar, A. Chandra, R. Varma, Grid interconnection of renewable energy sources at the distribution level with power-quality improvement features. IEEE Trans. Power Delivery **26** (1) (2011)
29. G.P. Zhang, A neural network ensemble method with jittered training data for time series forecasting. Inf. Sci. **177**, 5329–5346 (2007)
30. G.P. Zhang, Time series forecasting using a hybrid ARIMA and neural network model. Neurocomputing **50**, 159–175 (2003)
31. R. Adhikari, R.K. Agrawal, *An Introductory Study on Time Series Modeling and Forecasting*, pp. 1–67. https://arxiv.org/pdf/1302.6613
32. H. Chen, C.A. Canizares, A. Singh, ANN-based short-term load forecasting in electricity markets, in *Proceedings of the IEEE Power Engineering Society Transmission and Distribution Conference*, vol. 2 (2001), pp. 411–415
33. H. Park, Forecasting three-month treasury bills using ARIMA and GARCH models. Econ 930, Department of Economics, Kansas State University (1999)
34. J.W. Taylor, L.M. de Menezes, P.E. McSharry, A comparison of univariate methods for forecasting electricity demand up to a day ahead. Int. J. Forecast. **22**, 1–36 (2006)

Internet of Things—The Concept, Inherent Security Challenges and Recommended Solutions

Burhan Ul Islam Khan, Rashidah F. Olanrewaju, Farhat Anwar, Roohie Naaz Mir, Allama Oussama and Ahmad Zamani Bin Jusoh

Abstract The exponential growth observed as well as predicted in the development and deployment of the Internet of Things (IoT) based applications in every walk of our life brings forth the mandatory requirement of the secure communication system which is seamless yet effective in the highly heterogeneous and resource-constrained network. The core philosophy of the proposed research work is that existing cryptographic modelling will be required to be scaled down for its complexities by investigating the actual communication problems in security protocols between sensor nodes, Internet host, data centres, cloud clusters, virtual machines, etc. The inclusion of more operational actions while modelling security protocols is highly prioritised in the proposed research work. In this paper, the prime emphasis is (i) to establish secure pipelining using novel public key cryptography between the sensor network and Internet host, (ii) a robust authentication scheme considering both local and global IoT to offer better secure pervasiveness in its applications, and (iii) to apply optimisation towards key management techniques.

Keywords Authentication · Internet of Things · IoT security · Key management

1 Introduction

The technological advancement of hardware technology, sensors, actuators, software, middleware and firmware along with networking, communication, computing models and predictive methods provides possibilities of building an intelligent and automatic system of systems. One such vision is conceptualised in the recent past is the Internet of Things (IoT), where the physical world collaborates with intelligent ecosystem of

B. U. I. Khan (✉) · R. F. Olanrewaju · F. Anwar · A. Oussama · A. Z. B. Jusoh
Department of ECE, Kulliyyah of Engineering, International Islamic University Malaysia, Kualalumpur, Malaysia
e-mail: burhan.iium@gmail.com

R. N. Mir
Department of CSE, National Institute of Technology Srinagar, Hazratbal, Srinagar, Kashmir, India

M. Elhoseny and A. K. Singh (eds.), *Smart Network Inspired Paradigm and Approaches in IoT Applications*, https://doi.org/10.1007/978-981-13-8614-5_5

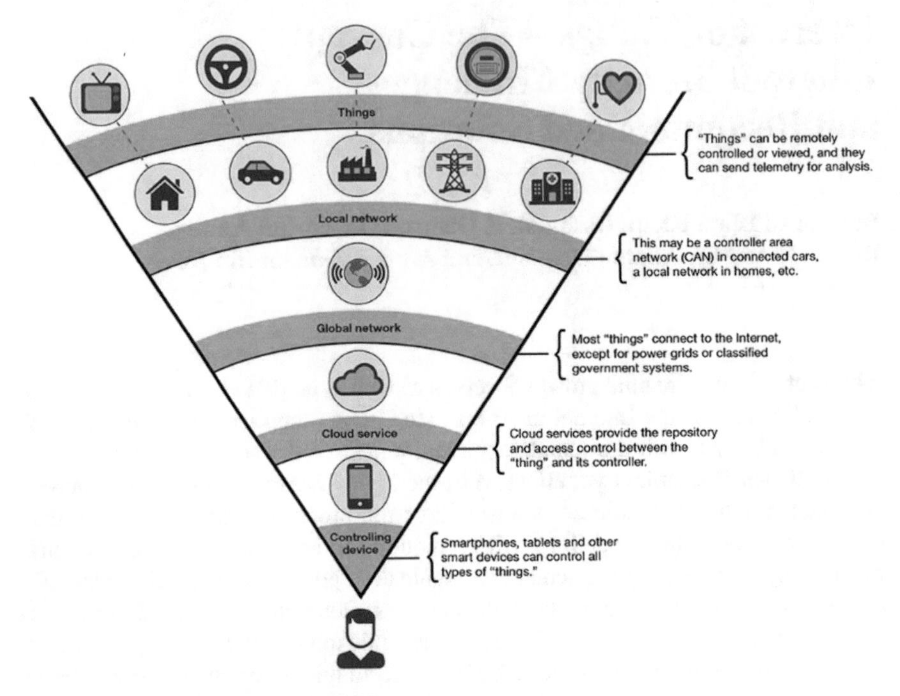

Fig. 1 IBM model for IoT (reproduced from [29])

the digital world. The IoT application may include smart logistic, smart grids, green building, smart transport, smart health care, smart homes, smart cities, etc. [3, 25, 27, 31, 97].

The Internet of Things is composed of a system of computing devices, people, animals or objects, digital and mechanical machines that are interrelated and have unique identifiers besides having the capability to transfer data over networks with no requirement of human-to-computer or human-to-human interaction. Figure 1 demonstrates the model of IoT as given by IBM [29]. The evolution of IoT was the result of the convergence of the Internet, microelectromechanical systems (MEMS), wireless technologies and micro-services. This convergence has led to removing the silo walls between Information Technology (IT) and Operational Technology (OT), thus paving the way for the analysis of unstructured machine-generated data to gain insights into driving improvements. Today, the practical applications of the Internet of Things technology are visible in numerous industries like transportation, health care, precision agriculture, energy and building management (shown in Fig. 2).

The increasing adoption of such applications leads the world to have a vast amount of the devices called IoT Node (Internet Node), which makes the vision of IoT possible ('IoT Sensor Node Block Diagram | Mouser', 2017) [42]. These innovative networks require abundant information as an input to make application realisation possible, and many a time this includes critical personal credentials such as

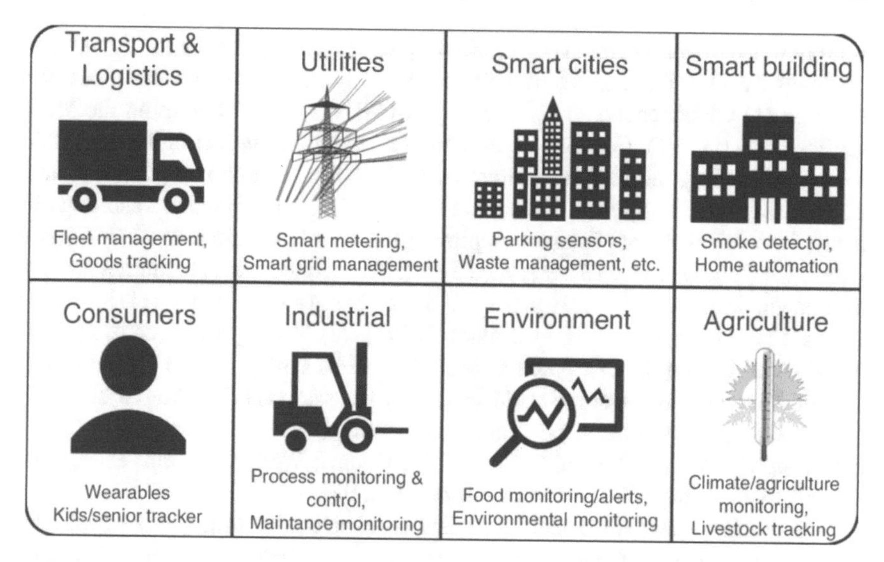

Fig. 2 Applications of IoT (reproduced from [11])

government-issued identification number, financial data, etc. [40, 72]. The threat of security grows exponentially O (n (IoT))2 with the increase of IoT Node. The possible attacks threats, and challenges concerning the security of any network depend upon the deployment architecture, and IoT is not an exception to this [44, 68]. Users, as well as enterprises, should be well-prepared for the large number of challenges posed by IoT. Some of the risks inherent in IoT together with the countermeasures have been enlisted below:

i. Disruptive cyber-attacks, like distributed denial-of-service attacks (DDoS), could prove detrimental to an organisation.
ii. To ward-off interruptions and potential operational failures in enterprise services, it is crucial to ensure the constant availability of IoT devices.
iii. Significant issues for an organisation include identification of sites in need of security controls and the implementation of adequate controls.
iv. In the IoT environment, discovering how IoT device vulnerabilities may be dispatched quickly and prioritising vulnerability patching are the next prominent challenges.

The relevant literature reveals that though there are technological improvements as well as IoT applications deployed for research and by industry but still as of now there is no standard architecture available [1, 14, 60], which poses many challenges and limits the effective performance of the IoT applications in real-time scenarios. Other related problems include (1) Communication bottlenecks, (2) Data Management, (3) Zero-entropy system, (4) Scalability, (5) Huge data collection and real-time processing of data, (6) Lack of standardisation, (7) Issues of interoperability, and finally last but not the least (8) Security and privacy [1, 4, 5, 45]. Though lately, there is an initiative towards global standardisation of security apparatus for the Internet of things by ITU-T ('Internet of Things', 2017) [41], but initially the focus was more on (1) Global Standard Initiatives (IoT-GSI), (2) Focus Group on the M2M service layer (ITU-T), (3) Joint Coordination Activity on Internet of Things (JCA-IoT), (4) Naming, numbering, addressing (ITU-T Study Group 2) ('Study Group 2 at a glance') [93], (5) Next Generation Network (NGN) architecture and requirements for services and applications employing tag-based identification (ITU-T Study Group 13) ('Study Group 13 at a glance') [90], (6) Architecture and requirements for multimedia information access prompted by tag-based identification (ITU-T Study Group 16) ('Study Group 16 at a glance') [91], (7) Testing architecture for tag-based identification (ITU-T Study Group 11) ('Study Group 11 at a glance') [89], (8) Global management of the radio-frequency spectrum (ITU-T Study Group 17) ('Study Group 17 at a glance') [92] and very recently (9) Security and privacy of tag-based applications (ITU-T Study Group 17) ('Council Working Groups and Expert Group', 2017; 'Study Group 17 at a glance') [23, 92].

One crucial aspect is the lack of unified architecture, as in the case of IoT, the constituent devices are heterogeneous over the wireless link and the current devices lack the capabilities to synchronise seamlessly with collocated devices employing different communication standards. Thus, these devices can only communicate with each other via IoT gateway nodes, hence causing the non-optimal use of wireless resources. Various efforts are set forth towards the unification of the IoT architecture [32], but the question to be investigated is, what kind of security threats or challenges could arise if there is a unification of architecture [54].

Therefore, a research focus is highly required to develop a unified architecture of secure framework solution to be applicable for most of the IoT applications with features of scalability, optimality, and robustness for effective deployment of IoT applications like smart cities, smart home, smart grids, smart transport systems, smart hospitals, etc. The designed security mechanisms need to ensure confidentiality, authentication, integrity and non-repudiation of the information flows across IoT application.

The remaining paper is organised into five sections where Sect. 2 presents the evolution of research in IoT concerning security aspects. Sections 3 and 4 discuss the significant limitations and the elaboration on the most critical security issues in IoT, respectively. The recommended solutions to overcome the challenges posed by IoT followed by the resulting consequences of bringing those into effect have been given in Sect. 5. Finally, the concluding remarks have been elucidated in Sect. 6.

2 Security-Based Research Evolution in IoT

This section discusses the general background to the exploration of various aspects of the security research evolution in the context of IoT. Also, it specifies the existing research efforts based on the three issues referred to in the introduction section, mainly (1) Secure channel between sensor motes and IoT Gateways, (2) Authentication and (3) Key management in IoT.

There exists specific archival survey work for security in IoT. These detailed surveys about the existing protocols and open issues can be found in [30, 62, 66, 75, 83, 85, 99, 101, 106]. The critical issues identified in connection to the current study of research work mainly comprise of: (i) How to develop security mechanism for establishing a secure channel between wireless sensor motes and IoT gateways (ii) Various mechanisms for authentication in order to ensure complete data confidentiality and data integrity for IoT-based applications, and (iii) Various key management schemes, which are widely suitable for IoT characteristics.

The security concerns are acute in the context of IoT after having some shreds of evidence of attacks such as (1) hacking of baby monitors, smart fridges, thermostats, drug fusion pumps, etc., and they can create severe havoc and losses [64]. Apart from device security, such problems also pose a threat to data security [46, 71, 109, 110]. In the coming years, IoT would get a permanent position within the healthcare and hygiene systems. Thus, future security will no longer linger around the information and commercial assets; however, our very lives would be at risk given the successful IoT security breaches [8]. The approach of security can be synchronised with the primary classification of the layered architecture of IoT. As per [57, 107], the typical layers include: (1) application layer, (2) network layer and (3) perception layer. Further, by introducing a service layer to the existing three-layer architecture, four-layer architecture, namely, Service-Oriented Architecture (SoA) based IoT architecture is formulated [95].

The classification of the security requirement in IoT can be broadly given in three categories (1) Security features of IoT which further includes: (i) Confidentiality, (ii) Integrity, (iii) Availability, (iv) Identification and Authentication, (v) Privacy and (vi) Trust. The second aspect (2) Layer-wise security challenges: wherein (i) Perception Layer: it includes (a) Node Capture Attacks, (b) Malicious Code Injection Attacks, (c) False Data Injection Attacks, (d) Replay Attacks, (e) Cryptanalysis Attacks and Side Channel Attacks, (f) Eavesdropping and Interference. In (ii) Network Layer: it includes (a) Denial-of-Service (DoS) Attacks, (b) Spoofing Attacks, (c) Sinkhole Attacks, (d) Wormhole Attacks, (e) Man-in-the-Middle Attack, (f) Routing Information Attacks, (g) Sybil Attacks, (h) Unauthorised Access. In (iii) Application Layer: it includes (a) Phishing Attack, (b) Malicious Virus/worm (c) Malicious Scripts, and finally, the (3) Privacy issues for consumers [54]. Figure 3, [106] describes the IoT security architecture.

A typical IoT possesses three characteristics, viz., (1) Comprehensive perception: It means the sensor nodes of the perception layer acquire the object information anytime and anywhere, (2) Reliable transmission: It indicates a safe and complete transmission of the information of objects via wired/wireless network to the data centre in real time, and (3) Intelligent processing: It implies that the middleware examines and deals with the information gathered before submission to the application terminal eventually [106].

Figure 4 shows layer-wise security provisioning in IoT; at the application layer, data security and application support security is the prime concern, whereas, at the

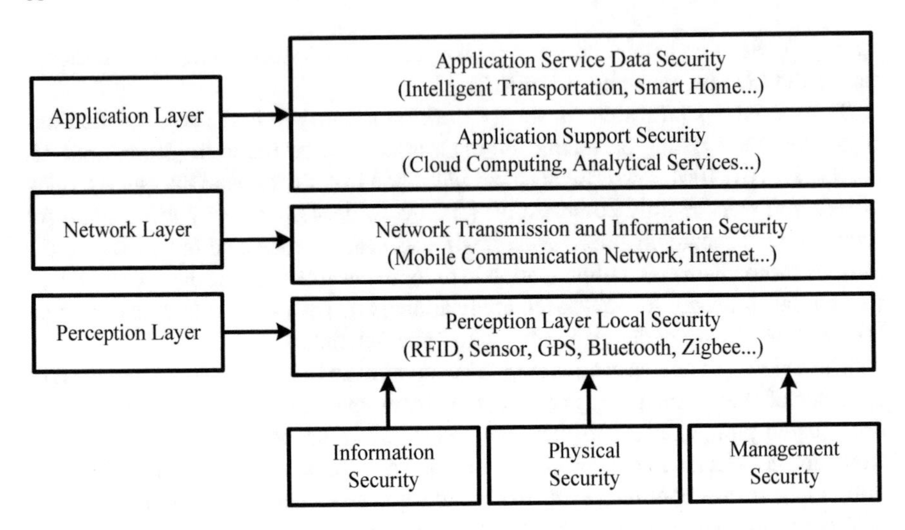

Fig. 3 Conceptual IoT security architecture

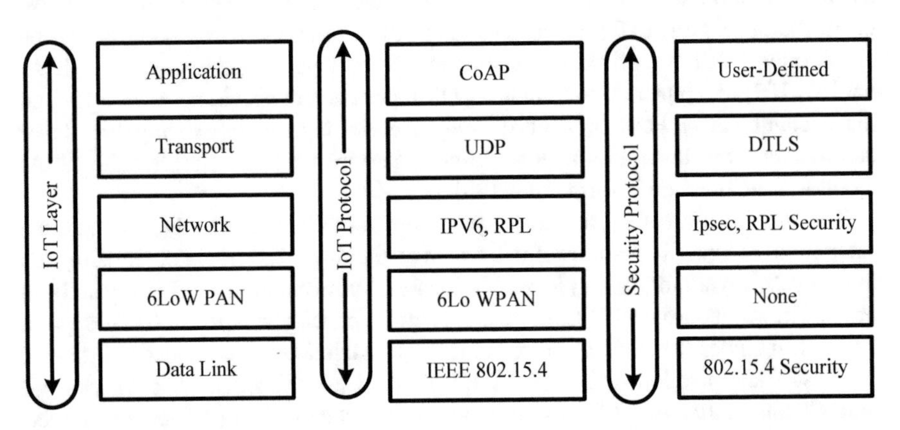

Fig. 4 Layer-wise standard security solutions

network layer, network transmission and information security are of prime interest; finally, at perception layer, local security needs to be considered [102]. The entire security paradigm for the framework design includes (1) Information security, (2) Physical security and finally (3) Management security.

The security requirements in IoT can be classified as (1) Communication security requirements, and (2) Data security requirements [83]. The communication security in IoT can be provided instead it can be said that the communication in IoT is protected by provisioning the desired security services including authentication, confidentiality, data integrity and access control. The description of different layers in IoT, existing IoT protocol and security provisioning which exists conventionally is shown in Fig. 4.

The state-of-the-art security at link layer includes IEEE 802.15.4 Link Layer security solution in the context of IoT, in it the underlying assumption is that every node in the communication path needs to be trusted [24] besides the use of a single pre-shared key throughout the communication. This design characteristic leads to the possibility of compromising the entire network even if one node is compromised. Furthermore, IEEE 802.15.4 cannot protect the acknowledgement messages regarding integrity or confidentiality. An opponent may thus forge acknowledgements, for which it only requires learning the sequence number of packets to assure that is sent in the clear, for conducting DoS attacks [30].

In the next layer of Low-Power Wireless Personal Area Network (6LoWPAN), IPv6 addressing mechanism is adopted as it supports commissioning, configuration management and network debugging. The Internet Engineering Task Force (IETF) initiatives of a task force for the support of IPv6 over IEEE 802.15.4. LoWPAN networks recommend (a) LoWPAN, with one edge router, (b) Ad hoc LoWPAN with no infrastructure and (c) LoWPAN with multiple edge routers which facilitates the authentication to establish connectivity without revealing identity of its neighbours, then it goes to the authentication, key generation and distribution process that makes it more vulnerable and thus there arises a need for more secure yet suitable for IoT authentication and key mechanisms.

In the network layer, primarily IP security and RPL security is implemented as the implementation platform for the IoT, and IP security (IPsec) is provisioned for end-to-end security along with authentication, confidentiality and integrity by the network layer as there is support for Constrained Application Protocol (CoAP), a transport layer protocol [86].

Although communication security has its critical requirements, at the same time, securing data which are generated from the participating IoT devices are equally important. The smaller size of the IoT devices poses unique constraints to secure them from the threats or security requirements which are related to the hardware. The multiple solutions existing are not synchronizable to the heterogeneous communication technologies; therefore, the concept of one solution to various problems is not adequate to proceed as the research line.

Towards the secure data storage issue, various solutions have been proposed in which one of the state-of-the-art works including Codo [7], which is a security extension of the file system in the Contiki OS [26] proposed for a secure communication and storage framework based on the 6LoWPAN/IPv6 protocols. The 6LoWPAN/IPv6 conceptualises the definition of IPsec as Encapsulating Security Payload (ESP), which provides the authentication and encryption of the transmitted data packets by utilising cryptographic methods. It uses a mechanism for including both header and data information in the cryptographic process. Further, to meet the compatibility of the communication standard, it needs to be stored into ESP using conventional mechanisms of key exchange which are designed for IPsec, that makes it heavy weighted and thus more research is required to make it comply with the architecture and constraints of IoT ecosystems.

There are working groups formed by IEEE and IETF towards standardisation and designing of new security and communications protocols which will play a significant role to enable future IoT applications. The prime objective of protocol designs is to provide optimal security to the communication to ensure the requirements of confidentiality, integrity, authentication, and non-repudiation of the information flow [17]. IoT communication security is generally addressed through communication protocol orientations only.

The other requirement towards security, which needs to be considered in the context of IoT is secure communication of the sensor or sensing devices with the Internet as some attacks that originated on the Internet like Denial of Service (DoS) may affect the availability and resilience [6].

There is also a requirement of a suitable security mechanism towards the normal working of IoT communication protocol such as fragmentation attacks at the 6LoWPAN at the adaptation layer [39].

Last but not the least, the significant security requirements include anonymity, privacy, trust and liability, that shall be vital for the social acceptance of the majority of future IoT applications using Internet-integrated sensing devices [13, 16, 69]. Various security schemes proposed by researchers have been analysed, and their limitations have been presented in Table 1.

Table 1 Summary of the findings

Author	Contribution	Limitations
Zhao et al. [105]	A lightweight mutual identity authentication scheme for IoT	• No concept of hierarchical accesses control • Consideration of general IoT • Heterogeneity within the network not taken into account
Ye et al. [104]	Efficient access control and authentication method for the perception layer of IoT	• Consideration of general IoT • Limited applicability • Heterogeneity among nodes at perception layer not considered • Provides forward secrecy only
Hu et al. [37]	Mutual identity authentication using modified elliptic mapping in authentication and key update mechanisms for multi-hop relay	• Not tested on open architecture/Standard architecture • Heterogeneity of the network not considered • No distinctions between global and local IoT • Absence of authorisation component
Hu et al. [73]	Capability Based Access Control (CBAC) and Elliptic Curve Cryptography based Mutual Authentication (EMA) model for securing authorisation	• No distinctions between global and local IoT • Heterogeneity of nodes not considered

(continued)

Table 1 (continued)

Author	Contribution	Limitations
Ma et al. [56]	Authentication protocol based on quantum key distribution using decoy-state method for heterogeneous IoT	• Authorization Component is missing • Consideration of general IoT
Barreto et al. [10]	IBC-based signcryption scheme	• Can't be used for heterogeneous communication
Li et al. [49]	PKI-based signcryption scheme	• Can't be used for heterogeneous communication
Sun and Li [94]	Heterogeneous signcryption scheme	• Not Secure against Insider attacks • High computational cost makes it unsuitable in WSN scenarios
Huang et al. [38]	Heterogeneous signcryption with provisions of Key privacy.	• High computational cost makes it unsuitable in WSN scenarios
Li and Xiong [50]	Heterogeneous Online/Offline based signcryption	• High offline storage requirement limits its applicability in PAN sensors • Usage of point multiplication limits the applicability of resource-constrained devices
Li et al. [51]	Heterogeneous Certificateless online/offline signcryption	• No provisions to accommodate the secure mobility of motes • Offline storage issue is still unsolved
Anand and Routray [2]	Narrowband IoT (NBIoT), preferably employed in healthcare applications owing to low-energy requirements	• No provision of real-time service and bandwidth inability
Hammi et al. [33]	Symmetric asynchronous One Time Password (OTP)/Encryption for Perception layer of IoT	• No performance analysis conducted with other schemes
Srinivas et al. [88]	Novel key agreement and authentication scheme for WSNs utilising biohashing	• Not secure against Blackhole and Wormhole attacks

3 Prominent Vulnerabilities in IoT

The Internet security protocols prevalent depend on a widely trusted and well-known suite of cryptographic algorithms like the Advanced Encryption Standard (AES) algorithm, Diffie–Hellman (DH) algorithm for key agreement, secure hash algorithms SHA1 and SHA256, and Rivest–Shamir–Adelman (RSA) algorithm. This suite is augmented by the evolving set of algorithms—Elliptic Curve Cryptography (ECC). Although its adoption is being slackened by Intellectual Property Rights

(IPR) issues, the recent IPR disclosures and RFC 6090 publication encourage its adoption.

All these cryptographic algorithms were intended to be used with the availability of significant resources like memory and processor speed. The application of such cryptographic methods to IoT is not precise, and further examination is required for ensuring that these can be implemented successfully in the limited processing power and memory available in IoT. In the initial development of IoT protocol, developers look for AES-GCM that is a combination of ECC algorithm suite, encryption and authentication. With the clarity of available resources on general IoT devices, researchers shall advance towards finding better cipher suites after agreeing that the algorithm suites are not optimal. It can cause apprehension among the developers since users having more common computing resources might break the easily implementable schemes. Irrespective of the device trace, a minimum of 112-bit level of security should be ensured for every cryptographic technique that is the present standard for constrained devices. However, it should not be assumed that the attacker has the same restrictions on resources.

It should be ensured that security features are available to early adopters as and when required. Thus, IoT protocol suites must specify a security solution that is optional but mandatory to implement. This shall make sure the availability of security in every implementation though it can be configured to be used only when required like in closed environments. It is expected that initial deployments may proceed by disabling the security but exploits leveraging those susceptibilities shall evolve in a short span. An informative example includes the experiences with small business and home wireless deployments in the form of the rapid discovery and exploitation of weak cryptography. A similar result can be seen if IoT is deployed without security.

Key management is the most challenging facet of cryptographic security. Various Internet protocols were deployed for manual key management, and it is unlikely to scale the number of IoT devices manually. Besides the significant number of devices, critical security shall be difficult to deploy due to limited user interfaces. Even though manual keying of devices can be done on initial deployment, but automatic rekeying post-deployment is also necessary.

Another critical issue in the present Internet includes credentialing of devices and users, and these problems are aggravated by the expected restrictions on user interfaces and the total number of devices. After carefully reviewing existing research work, it is found that the presence of distributed computing in cloud resources is the prime cause of security loopholes along with various other reported reasons [61]. The current research work towards addressing security problems in the cloud has witnessed highly diversified solutions as well as an evolution of the new genre of research techniques [70]. Security techniques combining manual and automatic schemes for initial deployment shall be required in the IoT. Particularly, protocol pairing like those based on Bluetooth security might be needed to be included in the deployment strategy. Nevertheless, it is envisioned that static keys shall only provide added support to initial deployment instead of being employed as traffic keys. Among these protocols, Leap-of-Faith technologies like those used in the IPSec profile of 'Better Than Nothing Security' (BTNS) might fill an important role.

Usability issues pose another challenge in the IoT. Irrespective of the type of device, it is imperative for the devices to be deployed or used easily. It shall be a big issue even regarding networking, and the provision of proper security may be the upcoming research topic.

Last but not least, issues concerning privacy are also substantial. Besides, the assaults on IoT protocol suites demand the establishment of a security boundary monitoring and constraining IoT devices. Thus, past technologies from the intelligence and military communities like network guards that were used for preventing information leakage shall be required again.

4 Critical Security Issues in IoT

4.1 Secure Communication Issue Between WSN and IoT

The consideration of the resource constraints of the sensor motes and the scaling factors of the IoT make the protocols employed for the Internet unsuitable for the desired requirements of the communication and security among WSN and IoT. Technological solutions need to be devised in line with the characteristics and constraints of low-rate wireless communications and low-energy sensing devices. Such features have also shaped former application designs using Wireless Sensor Networks (WSN) segregated from the Internet, and abundant research proposals on security mechanisms [19]. The approach towards the development of such new designs is the provision of assured interoperability with traditional standards and secure and seamless communication between the sensing device and with IoT ecosystem especially Internet entities as future distributed IoT-based applications. In a nutshell, it can be said that if the sensor data needs to be made available globally, then WSN requires being an integral part of IoT and the integration of the WSN and the Internet includes three approaches as shown in Fig. 5 [81].

The security challenge for establishing a secure communication between sensor node and the IoT or Internet host is to meet the authentication as well as confidentiality with a contrast situation of collaboration among resource constraints of WSN in terms of power and memory along with the Internet host of comparatively abundant or higher computational capacity and memory including storage and buffer. Designing such a secure communication method or process to be suitable for such characteristics poses its challenges as it has to cope with multiple challenges which more significantly include: (1) WSN-specific security challenges and (2) Integration security challenges.

Towards the WSN-specific security challenges, the bare minimum requirements of security properties for the underlying core functionalities require the usage of cryptographic mechanism and key management with the functional support of self-configurability. The contemporary sensors are capable of the functioning of Symmetric Key Cryptography (SKC), hashing functions and Public-Key Cryptography

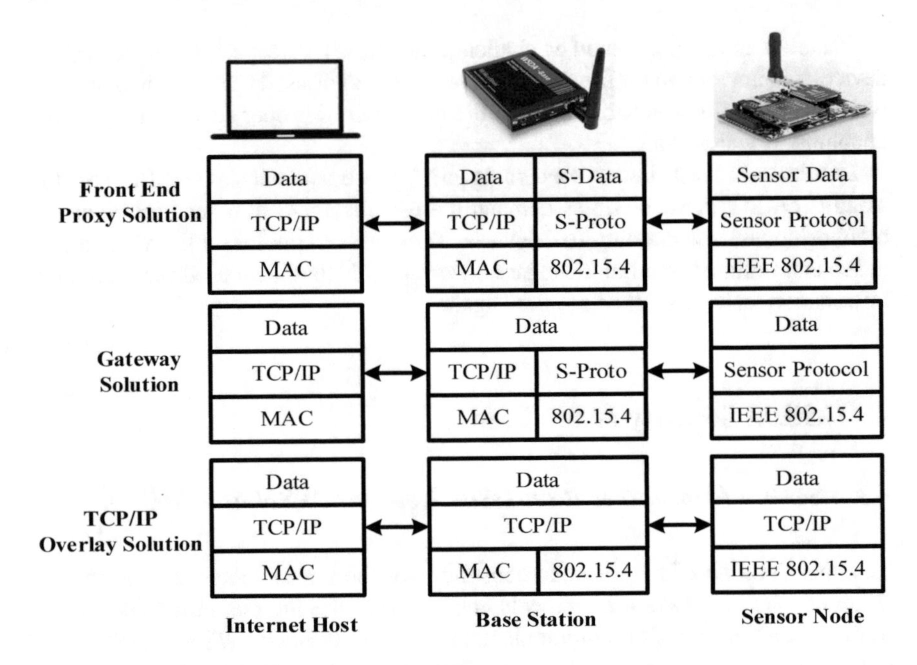

Fig. 5 Integration strategy between WSN and internet

(PKC); by default, IEEE 802.15.4 supports AES-128. There exist some related works where approximately 8 kilobyte of Read-Only-Memory (ROM), 300 bytes of Random Access Memory (RAM) by ES-128 implementation was used at the same time. Methods like block ciphers and stream ciphers such as Skipjack and RC4 demonstrate the small use of memory to the extension of approximately 2600 and 428 bytes, respectively [20, 48]. The traditional assumption that WSN cannot implement Public-Key Cryptography (PKC) was contested by the adoption of Elliptical Curve Cryptography (ECC) with the capabilities of encryption, decryption, signing, verifying and key establishment as in a resource-constrained sensor network but it too suffers challenges of larger memory and computational overheads [55]. Further, the implementation of hash functions like SHA-1 is realised with merely 3 kB of ROM.

Whereas the complexities of the WSN application and its scalability may exhaust the memory and implementing such ECC and hashing mechanism ceases its utility, so novel key management mechanisms exploring distributed link layer keys can provision seamless and secure integration of WSN with IoT/Internet with routing, aggregation and time synchronisation [82].

There exists a hybrid method of signature and encryption in a single logical way called Signcryption, which is comparably lower in computational cost if the sequential first signature, then encryption approach is used and has the potential to achieve the goal of confidentiality, integrity, authentication and non-repudiation at a lower cost.

Some significant Signcryption schemes in the Public Key Infrastructure (PKI) have been proposed by [9, 28, 28, 58]. However, the PKI is not suitable as the certificate management is resource hungry to facilitate its operations of distribution, storage and revocation. Therefore, to minimise the encumbrance effect of resource consumption of certificate methods, there exist some Signcryption schemes in the identity-based cryptography (IBC), which are [12, 21, 43, 53]. In contrast to the PKI, the IBC poses an advantage of eliminating the public key which is computed from the identity information of the user itself and the private key generator (PKG) as a trusted third party produces the private key for all the participating nodes, so the public key is authenticated without the verification of public key of the certificate. With respective merits and limitations of PKI and IBC, these Signcryption schemes are not adoptable to the heterogeneous communication.

The objective of achieving a secure communication channel between a heterogeneous system of the sensor node to IoT host, for the provision of end-to-end confidentiality, authentication, integrity and non-repudiation services in an optimal computational cost remains an open research issue.

4.2 Authentication Issue in the Internet of Things

In order to understand the authentication scheme for IoT, initially it requires to understand the underlying security architecture and adopted solutions, which includes the works of [22, 35, 47, 59, 67, 80], along with the study of secure communication and networking mechanisms [34, 77, 96, 100], cryptographic algorithms [18, 78, 80, 98, 103, 105] and finally the application security solutions [52, 87, 108]. Contemporary research works primarily refer to three aspects: application security, network security and system security.

The conventional mechanisms based on cryptology for identity management, trust-based frameworks, fault tolerance, privacy preservation face many challenges in the context of IoT security because of their heterogeneous network characteristics [80].

Various researchers have worked towards achieving the trusted and secure communication in a heterogeneous network environment; one of such work was by [47], where identity and authentication management in the future Internet is studied but in the context of online service payment. Further, one of the studies towards the IP-based IoT was [35], in which they have analysed and elaborated the applicability and limitations of the conventional security architecture of the IP networks and advocated the consideration of security architecture, node security model, etc., in security solutions with only a resource-constrained heterogeneous communication. Furthermore, a security architecture to ensure secure network and key management by combining the Multimedia Internet Keying Protocol (MIKP) Host Identity Protocol (HIP) is proposed by [59].

After the system security, the researchers focussed on network security, where initially the security requirements towards the IoT system with RFID systems were studied by considering core challenges such as privacy, data integrity, application integrity, security standardisation, etc., and recommended to enhance it to suit for the universal security [34]. Other works towards the mobile security protocol, integration of RFID with IP network, trust-based cryptographic and authentication in WSN, the combination of Datagram Transport Layer Security (DTLS), Constrained Application Protocol (CoAP) in order to provide secure transmission in IoT [15, 77, 96, 100].

A paradigm shift towards authentication protocols has taken place in the domain of IoT security. Yao et al. [103] designed a message authentication code (MAC) for a small-sized IoT application. Roman et al. [80] have focused on developing key management mechanism among remote devices by evaluating the applicability of PKI and pre-shared keys for sensor nodes and IoT context. The use of bilinear mapping is evidenced in a work by [78] as attribute-based access control model and achieved anonymous access and minimisation of consumption of resources because of the reduced control message mechanism. Various other approaches like fuzzy reputation-based trust management model (TRM-IoT), unknown authentication protocol to achieve the trade-off between anonymity and certification is proposed by [18, 98]. Further, [105] introduced a mutual authentication scheme using SHA and Elliptic Curve Cryptography (ECC) to establish an asymmetric authentication scheme with the aim of achieving optimal computation cost and communication overhead.

Finally, towards the application security, a traffic security architecture for IoT for heterogeneous communication is introduced into the security architecture by [108]. Li et al. [52], presents a smart community model for IoT application with security considerations by filtering the malicious traffic emphasising the need for intrusion detection, cooperative authentication and unreliable node detection. Layer security for smart grid was proposed by [87], where again a layer security approach is established to evaluate the security risks for the power applications by emphasising on encryption, authentication, access control, etc.

To have a unified security architecture framework, some defined architecture needs to be modified to the global architecture and achieve the goal of authentication; one such architecture U2IoT could be generalised to suit this purpose.

4.3 Key Management Issue in the Internet of Things

The effort towards addressing security and privacy challenges for the ever-growing interconnectivity of billions of devices is on. Many initial solutions towards secure communication among the resource-constrained devices in IoT ecosystem have been tried to devise using light weighted IPsec, Datagram transport layer security (DTLS) and link layer security [39, 76], respectively, along with some network security in IoT such as intrusion detection system, data protection in nodes and standardisation towards security in IoT.

Various working committees by IETF and IEEE are being set up including DICE working group 1 with the aim of adopting DTLS into the constrained environment like

IoT [84]. Another goal of having a standardised way of authenticating and authorising access to a restricted node is an essential requirement for which a working group ACE -2 was set up by IETF in June 2014.

The perspective definition of IoT is also a machine to machine (M2M) communication; an Open Mobile Appliance (OMA) initiated a necessary draft from CoAP and DTLS along with an additional security binding CoAP over SMS. The original draft lacks the mechanism of securing the channel due to the absence of pre-established credentials.

The authentication protocol Kerberos [65] is the mother of all the authentication protocols where two nodes alien to each other create a shared key; the requirement in the case of IoT is to integrate DTLS. In this direction, [36] proposed an authorisation and authentication framework for smart objects to establish keys for DTLS. Further, [74] has designed a framework for access control and authentication in the CoAP-based Internet of Things [74]. Their framework leverages Kerberos [65] and RADIUS [79] to provide authentication and access control. But for the constraint devices, it is too heavy; so, a process and evolution of lightweight key management schemes are needed to secure the communication of IoT.

5 Recommended Solutions

The IoT application framework encompasses highly heterogeneous devices, computing migration among sensor to IoT gateways, gateways to fog/edge computing to cloud then back to the user as well as other alarming components. It was found from the literature that security provisioning of such a network is in nascent stages to the provision of a unified architecture to have seamless security along with the application and network prospects, though the shortfalls in the initiatives from IEEE and IETF have more contribution towards it. In addition, the existing researchers are far from achieving the real threshold of real-time performance possibilities in terms of lower computational cost (low space and time complexity), conformable security, lower key size usage, smaller size of ciphertext, reduced memory overheads, resiliency against potential threats, faster algorithm processing time (faster response time) and lower network/communication cost to ensure lighter security mechanism. Therefore, some solutions have been presented in this paper to tackle the most significant security issues in IoT that are given below.

5.1 Secure Transmission Between Sensor and IoT

The original design principle is based on the fact that public key cryptography has good supportability among the sensor nodes [63]. This part of the study considers that sensor motes are transmitters and Internet gateway (IoT) is the receiver. The duo has structural and computational differences; thus, two different forms of security

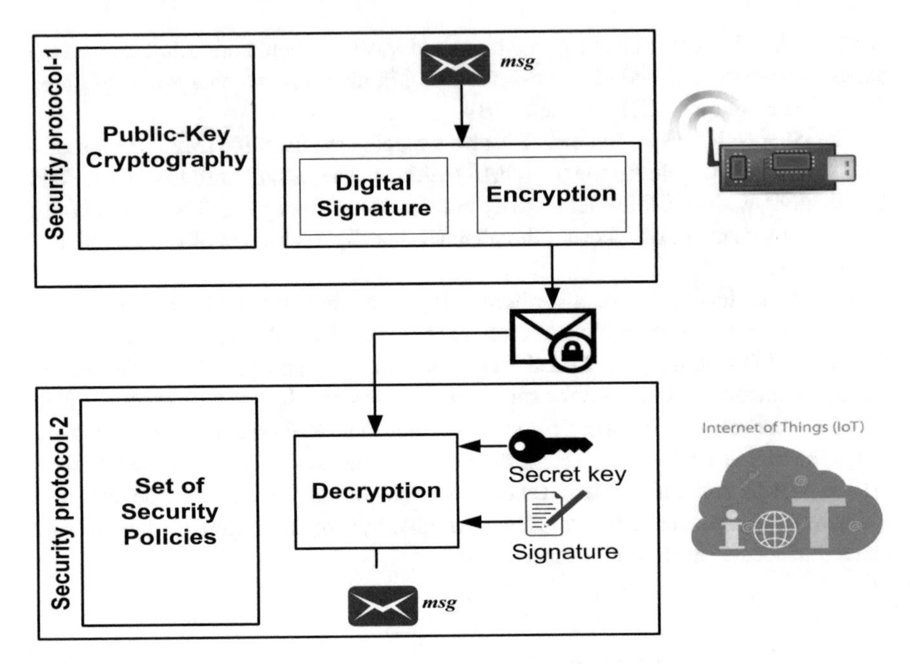

Fig. 6 A tentative schema for secure transmission between sensor and IoT

protocols will be formulated, one exclusively for sensor nodes within the WSN and other for the Internet gateway. The transmitter (sensor node) will capture their targeted data and then it will be subjected to the ciphering process using the digital signature as well as encryption. The encrypted data will be forwarded to the Internet host. The tentative schematic diagram is shown in Fig. 6.

Two algorithms will assist the proposed mechanism. The first algorithm will be responsible for the initial set of security parameters using new public key cryptography. It also formulates the master key arbitrarily to compute the public key (master). A key generation process is then carried out separately for both the security protocols. The first security protocol will choose to select a specific attribute (to be investigated) to its public-key encryption that is also responsible for computing the secret key.

On the other hand, the second security protocol will be more like a set of security policies as well as roles. In this protocol, the secret key will be computed based on the selection of random numbers by the receiver node (sensor mote in IoT). For efficiency of the cryptosystem, the proposed study will perform the implementation of encryptions and digital signature both in a single logical step (signcryption). To further strengthen the protocol, it needs to explore the possibility of secure communication between different sensor nodes connected directly/indirectly to IoT gateways; i.e. when IoT gateway is/is not the next immediate host for a particular sensor mote. Finally, the decryption will be carried out considering the secret key as well as the signature.

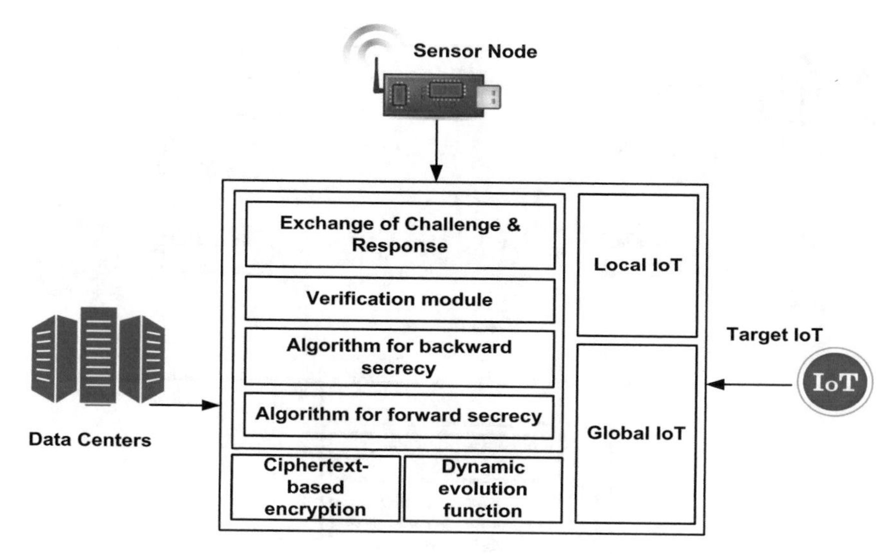

Fig. 7 Tentative schema for robust authentication in IoT

The desired outcome of the application of the solution recommended shall be as follows:

- Lower computational cost (low space and time complexity).
- Higher security.
- Optimal key size usage.
- Optimal size of ciphertext.

5.2 Robust Authentication in IoT

The main agenda of this part of the research work is to perform secure modelling of local and global IoT. Local IoT would mean a network with a single application while global IoT would mean forming the centralised network with multiple (and diverse) applications. Initially, simple modelling will be carried out by defining actors in local IoT (data centres, target IoT, sensors) and global IoT (the network of all the datacentres). The tentative research methodology is shown in Fig. 7.

A novel authentication mechanism will be presented among all the possible actors residing within the domain of local and global IoT by incorporating exchanging of challenge and response-based approach. A sophisticated verification module will be designed validating the legitimacy of the requestor node. Two distinct algorithms for forward and backward secrecy will be formulated to safeguard the presented authentication framework for any possible threats. The system will use ciphertext-based encryption and novel dynamic evolution function to strengthen the authentication mechanism further.

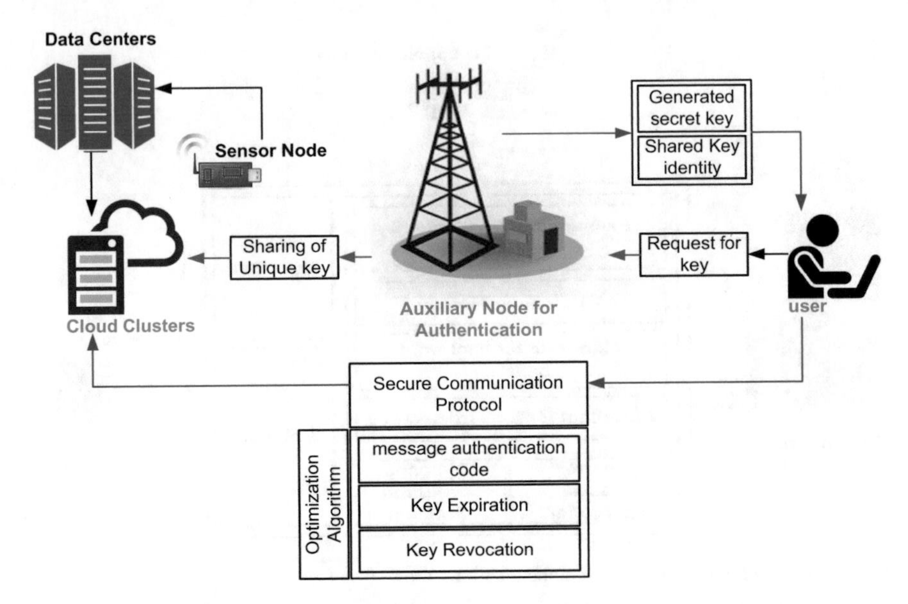

Fig. 8 Tentative schema for enhancing key management in IoT

The desired outcome of the application of this solution shall be

- Faster algorithm processing time (faster response time).
- Minimal communication overhead.
- Hierarchical access control.

5.3 *Enhancing Key Management in IoT*

The proposed work will consider an additional actor, i.e. an auxiliary node between cloud clusters and user. The cloud cluster is geographically spread and highly distributed that is assumed to offer resources requested by the user only with the help of auxiliary authenticator node. The multiple authenticator node shares a unique secret key with the highly distributed cloud clusters (that could also be multiple in numbers). The user will be required to request for a secret key to the auxiliary node, which in return offers a generated secret key and shared key identity to the user. This information of keys will be employed by the user to establish secure communication with the cloud clusters. Figure 8 shows the tentative schematic diagram of proposed work.

The optimisation algorithm will be created using message authentication code along with specifications of key expiry as well as revocation information. This work will use the framework of prior phases to implement robust authentication among all the actors involved in IoT-based communication.

The desired outcome of applying this solution shall be:

- Resilience against potential threats like replay attacks, DOS, etc.
- Scalable solution.
- Compliant with time and space complexity and will bridge the trade-off between communication and computation.

6 Conclusion

The research conducted concludes that the security challenges posed by the IoT are daunting. As a result, it is the need of the hour that security features mandatory to implement should be included in the early IoT protocols, even though such features extend the potentials of those devices. Furthermore, the automation in key management is a challenge as well but avoiding the reliance on pre-shared keys is more critical for IoT protocols. Another significant problem includes the registration or credentialing of devices, and a possible solution set can be the well-understood protocol pairing. Other than these, privacy issues may motivate the adoption of newer technologies that have been designed for the prevention of information leakage in intelligence/military environments. Three schematic methodologies have been given in this study to tackle these issues in IoT.

Acknowledgements This work was partially supported by the Ministry of Higher Education Malaysia (Kementerian Pendidikan Tinggi) under Research Initiative Grant Scheme number: **RIGS16-334-0498**.

References

1. S.A. Al-Qaseemi, H.A. Almulhim, M.F. Almulhim, S.R. Chaudhry, IoT architecture challenges and issues: lack of standardization, in *Future Technologies Conference (FTC)* (IEEE, 2016), pp. 731–738
2. S. Anand, S.K. Routray, Issues and challenges in healthcare narrowband IoT, in *International Conference on Inventive Communication and Computational Technologies (ICICCT)* (IEEE, 2017), pp. 486–489
3. V. Angelakis, E. Tragos, H. Pöhls, A. Kapovits, A. Bassi, *Designing, Developing, and Facilitating Smart Cities* (Springer, Berlin, 2017)
4. F. Anwar, M.H. Masud, B.U.I. Khan, R.F. Olanrewaju, S.A. Latif, Bandwidth allocation policy using the game theory model in heterogeneous wireless networks. IPASJ Int. J. Inf. Technol. (IIJIT) **6**(7), 1–8 (2018)
5. F. Anwar, M.H. Masud, B.U.I. Khan, R.F. Olanrewaju, S.A. Latif, Game theory for resource allocation in heterogeneous wireless networks-a review. Indones. J. Electr. Eng. Comput. Sci. **12**(2), 843–851 (2018)
6. A. Arış, S.F. Oktuğ, S.B. Yalçın, Internet-of-Things security: denial of service attacks, in *Signal Processing and Communications Applications Conference (SIU)* (IEEE, 2015), pp. 903–906

7. I.E. Bagci, M.R. Pourmirza, S. Raza, U. Roedig, T. Voigt, Codo: confidential data storage for wireless sensor networks, in *9th International Conference on Mobile Adhoc and Sensor Systems (MASS)* (IEEE, 2012), pp. 1–6
8. A. Banafa, Three major challenges facing IoT, in *IEEE Internet of Things* (2017). http://iot.ieee.org/newsletter/march-2017/three-major-challenges-facing-iot. Accessed 7 Feb 2018
9. F. Bao, R.H. Deng, W. Mao, Efficient and practical fair exchange protocols with off-line TTP, in *Symposium on Security and Privacy, Proceedings* (IEEE, 1998), pp. 77–85
10. P.S. Barreto, B. Libert, N. McCullagh, J.J. Quisquater, Efficient and provably-secure identity-based signatures and signcryption from bilinear maps, in *International Conference on the Theory and Application of Cryptology and Information Security* (Springer, Berlin, 2005), pp. 515–532
11. A.A. Boulogeorgos, P.D. Diamantoulakis, G.K. Karagiannidis, Low power wide area networks (LPWANs) for internet of things (IoT) applications: research challenges and future trends (2016). arXiv:1611.07449
12. X. Boyen, Multipurpose identity-based signcryption, in *Annual International Cryptology Conference* (Springer, Berlin, 2003), pp. 383–399
13. I. Butun, M. Erol-Kantarci, B. Kantarci, H. Song, Cloud-centric multi-level authentication as a service for secure public safety device networks. IEEE Commun. Mag. **54**(4), 47–53 (2016)
14. E. Cavalcante, M.P. Alves, T. Batista, F.C. Delicato, P.F. Pires, An analysis of reference architectures for the internet of things, in *Proceedings of the 1st International Workshop on Exploring Component-based Techniques for Constructing Reference Architectures* (ACM, 2003), pp. 13–16
15. K.D. Chang, J.L. Chen, A survey of trust management in WSNs, internet of things and future internet. KSII Trans. Internet Inf. Syst. **6**(1), 5–23 (2012)
16. D. Chasaki, C. Mansour, Security challenges in the internet of things. Int. J. Space-Based Situated Comput. **5**(3), 141–149 (2015)
17. D. Chasaki, C. Mansour, Selective encryption of video transmissions over multi-hop wireless networks, in *Symposium on Computers and Communication (ISCC)* (IEEE, 2014), pp. 1–5
18. D. Chen, G. Chang, D. Sun, J. Li, J. Jia, X. Wang, TRM-IoT: a trust management model based on fuzzy reputation for internet of things. Comput. Sci. Inf. Syst. **8**(4), 1207–1228 (2011)
19. X. Chen, K. Makki, K. Yen, N. Pissinou, Sensor network security: a survey. IEEE Commun. Surv. Tutor. **11**(2), 52–73 (2009)
20. K.J. Choi, J.I. Song, Investigation of feasible cryptographic algorithms for wireless sensor network, in *The 8th International Conference on Advanced Communication Technology, ICACT*, vol. 2 (IEEE, 2006), pp. 1379–1381
21. S.S. Chow, S.M. Yiu, L.C. Hui, K.P. Chow, Efficient forward and provably secure ID-based signcryption scheme with public verifiability and public ciphertext authenticity, in *International Conference on Information Security and Cryptology* (Springer, Berlin, 2003), pp. 352–369
22. M. Conti, A. Dehghantanha, K. Franke, S. Watson, Internet of Things security and forensics: challenges and opportunities. Futur. Gener. Comput. Syst. **78**, 544–546
23. Council Working Groups and Expert Group. http://www.itu.int/en/council/Pages/groups.aspx. Accessed 7 Feb 2018
24. A. Cui, S.J. Stolfo, A quantitative analysis of the insecurity of embedded network devices: results of a wide-area scan, in *Proceedings of the 26th Annual Computer Security Applications Conference* (ACM, 2010), pp. 97–106
25. H.S. Dhillon, H. Huang, H. Viswanathan, Wide-area wireless communication challenges for the Internet of Things. IEEE Commun. Mag. **55**(2), 168–174 (2017)
26. A. Dunkels, B. Gronvall, T. Voigt, Contiki-a lightweight and flexible operating system for tiny networked sensors, in *29th Annual IEEE International Conference on Local Computer Networks* (IEEE, 2004), pp. 455–462
27. S. Feng, P. Setoodeh, S. Haykin, Smart home: cognitive interactive people-centric Internet of Things. IEEE Commun. Mag. **55**(2), 34–39 (2017)

28. C. Gamage, J. Leiwo, Y. Zheng, Encrypted message authentication by firewalls, in *International Workshop on Public Key Cryptography* (Springer, Berlin, 1999), pp. 69–81
29. A. Gerber, Simplify the development of your IoT solutions with IoT architectures: strategies for creating scalable, flexible, and robust IoT solutions (2017). https://www.ibm.com/developerworks/library/iot-lp201-iot-architectures/. Accessed 8 Feb 2018
30. J. Granjal, E. Monteiro, J.S. Silva, Security for the internet of things: a survey of existing protocols and open research issues. IEEE Commun. Surv. Tutor. **17**(3), 1294–1312 (2015)
31. S. Greengard, The Internet of Things. (MIT Press, 2015)
32. Y. Guo, H. Zhu, L. Yang, Smart service system (SSS): a novel architecture enabling coordination of heterogeneous networking technologies and devices for internet of things. China Commun. **14**(3), 130–144 (2017)
33. M.T. Hammi, E. Livolant, P. Bellot, A. Serhrouchni, P. Minet, A lightweight mutual authentication protocol for the IoT, in *International Conference on Mobile and Wireless Technology* (Springer, Singapore, 2017), pp. 3–12
34. G.P. Hancke, K. Markantonakis, K.E. Mayes, Security challenges for user-oriented RFID applications within the Internet of things. J. Internet Technol. **11**(3), 307–313 (2010)
35. T. Heer, O. Garcia-Morchon, R. Hummen, S.L. Keoh, S.S. Kumar, K. Wehrle, Security challenges in the IP-based Internet of Things. Wireless Pers. Commun. **61**(3), 527–542 (2011)
36. J.L. Hernandez-Ramos, M.P. Pawlowski, A.J. Jara, A.F. Skarmeta, L. Ladid, Toward a lightweight authentication and authorization framework for smart objects. IEEE J. Sel. Areas Commun. **33**(4), 690–702 (2015)
37. T. Hu, J. Wang, G. Zhao, X. Long, An improved mutual authentication and key update scheme for multi-hop relay in Internet of Things, in *7th IEEE Conference on Industrial Electronics and Applications (ICIEA)* (IEEE, 2012), pp. 1024–1029
38. Q. Huang, D.S. Wong, G. Yang, Heterogeneous signcryption with key privacy. Comput. J. **54**(4), 525–536 (2011)
39. R. Hummen, J. Hiller, H. Wirtz, M. Henze, H. Shafagh, K. Wehrle, 6LoWPAN fragmentation attacks and mitigation mechanisms, in *Proceedings of The Sixth ACM Conference on Security and Privacy in Wireless and Mobile Networks* (ACM, 2013), pp. 55–66
40. Internet of things research study 2015 report. https://www.hpe.com/h20195/V4/Getdocument.aspx?docname=4AA5-4759ENW. Accessed 7 Feb 2018
41. Internet of Things. http://www.itu.int/en/ITU-T/techwatch/Pages/internetofthings.aspx. Accessed 8 Feb 2018
42. IoT Sensor Node Block Diagram | Mouser. https://www.mouser.in/applications/internet-of-things-block-diagram/. Accessed 8 Feb 2018
43. H.J. Jo, J.H. Paik, D.H. Lee, Efficient privacy-preserving authentication in wireless mobile networks. IEEE Trans. Mob. Comput. **13**(7), 1469–1481 (2014)
44. B.U.I. Khan, R.F. Olanrewaju, F. Anwar, R.N. Mir, A.R. Najeeb, A critical insight into the effectiveness of research methods evolved to secure IoT ecosystem. Int. J. Inf. Comput. Secur. (2018) (In press)
45. B.U.I. Khan, A.M. Baba, R.F. Olanrewaju, S.A. Lone, N.F. Zulkurnain, SSM: Secure-Split-Merge data distribution in cloud infrastructure, in *IEEE Conference on Open Systems (ICOS)* (IEEE, 2015), pp. 40–45
46. B.U.I. Khan, R.F. Olanrewaju, M.H. Habaebi, Malicious behaviour of node and its significant security techniques in MANET-A review. Aust. J. Basic Appl. Sci. **7**(12), 286–293 (2013)
47. K. Lampropoulos, S. Denazis, Identity management directions in future internet. IEEE Commun. Mag. **49**(12), 74–83 (2011)
48. Y.W. Law, J. Doumen, P. Hartel, Survey and benchmark of block ciphers for wireless sensor networks. ACM Trans. Sens. Netw. (TOSN) **2**(1), 65–93 (2006)
49. C.K. Li, G. Yang, D.S. Wong, X. Deng, S.S. Chow, An efficient signcryption scheme with key privacy and its extension to ring signcryption. J. Comput. Secur. **18**(3), 451–473 (2010)
50. F. Li, P. Xiong, Practical secure communication for integrating wireless sensor networks into the internet of things. IEEE Sens. J. **13**(10), 3677–3684 (2013)

51. F. Li, Y. Han, C. Jin, Practical signcryption for secure communication of wireless sensor networks. Wireless Pers. Commun. **89**(4), 1391–1412 (2016)
52. X. Li, R. Lu, X. Liang, X. Shen, J. Chen, X. Lin, Smart community: an internet of things application. IEEE Commun. Mag. **49**(11), 68–75 (2011)
53. B. Libert, J.J. Quisquater, Identity based undeniable signatures, in *Cryptographers' track at the RSA conference* (Springer, Berlin, 2004), pp. 112–125
54. J. Lin, W. Yu, N. Zhang, X. Yang, H. Zhang, W. Zhao, A survey on internet of things: architecture, enabling technologies, security and privacy, and applications. IEEE Internet Things J. **4**(5), 1125–1142 (2017)
55. A. Liu, N. TinyECC, A configurable library for elliptic curve cryptography in wireless sensor networks 2008, in *Proceedings of the 7th International Conference on Information Processing in Sensor Networks* (IEEE Computer Society, Washington DC, 2008), pp. 245–256
56. H. Ma, B. Chen, An authentication protocol based on quantum key distribution using decoy-state method for heterogeneous IoT. Wireless Pers. Commun. **91**(3), 1335–1344 (2016)
57. R. Mahmoud, T. Yousuf, F. Aloul, I. Zualkernan, Internet of things (IoT) security: current status, challenges and prospective measures, in *10th International Conference for Internet Technology and Secured Transactions (ICITST)* (IEEE, 2015), pp. 336–341
58. J. Malone-Lee, W. Mao, Two birds one stone: signcryption using RSA, in *Cryptographers' Track at the RSA Conference* (Springer, Berlin, 2003), pp. 211–226
59. F.V. Meca, J.H. Ziegeldorf, P.M. Sanchez, O.G. Morchon, S.S. Kumar, S.L. Keoh, HIP security architecture for the IP-based internet of things, in *27th International Conference on Advanced Information Networking and Applications Workshops (WAINA)* (IEEE, 2013), pp. 1331–1336
60. A. Meddeb, Internet of things standards: who stands out from the crowd? IEEE Commun. Mag. **54**(7), 40–47 (2016)
61. M.S. Mir, B. Suhaimi, M. Adam, B.U.I. Khan, M.M.U.I. Mattoo, R.F. Olanrewaju, Critical security challenges in cloud computing environment: an appraisal. J. Theor. Appl. Inf. Technol. **95**(10), 2234–2248 (2017)
62. A. Mosenia, N.K. Jha, A comprehensive study of security of internet-of-things. IEEE Trans. Emerg. Top. Comput. **5**(4), 586–602 (2017)
63. I. Nadir, W.K. Zegeye, F. Moazzami, Y. Astatke, Establishing symmetric pairwise-keys using public-key cryptography in Wireless Sensor Networks (WSN), in *IEEE Annual Ubiquitous Computing, Electronics & Mobile Communication Conference (UEMCON)* (IEEE, 2016), pp. 1–6
64. K. Narayanan, Addressing the challenges facing IoT adoption. Microw. J. **60**(1), 110–118 (2017)
65. C. Neuman, J. Kohl RFC 4120: the Kerberos network authentication service (V5) 2005 (2015)
66. K.T. Nguyen, M. Laurent, N. Oualha, Survey on secure communication protocols for the Internet of Things. Ad Hoc Netw. **32**, 17–31 (2015)
67. H. Ning, H. Liu, L. Yang, Cyber-entity security in the Internet of things. Computer **46**(4), 46–53 (2013)
68. R.F. Olanrewaju, B.U.I. Khan, A.L. Mechraoui, Game theory based probabilistic approach to detect misbehaving nodes in ad-hoc networks, in *Proceedings of the 2nd IEEE International Conference on Intelligent Systems Engineering (ICISE)*, Kuala Lumpur, Malaysia (2018)
69. R.F. Olanrewaju, B.U.I. Khan, M.M. Mattoo, F. Anwar, A.N. Nordin, R.N. Mir, Z. Noor, Adoption of cloud computing in higher learning institutions: a systematic review. Indian J. Sci. Technol. **10**(36), 1–9 (2017)
70. R.F. Olanrewaju, B.U.I. Khan, A. Baba, R.N. Mir, S.A. Lone, RFDA: reliable framework for data administration based on split-merge policy, in *SAI Computing Conference (SAI)* (IEEE, 2016), pp. 545–552
71. R.F. Olanrewaju, B.U.I. Khan, R.N. Mir, A. Shah, Behaviour visualization for malicious-attacker node collusion in MANET based on probabilistic approach. Am. J. Comput. Sci. Eng. **2**(3), 10–19 (2015)
72. V. Oleshchuk, Internet of things and privacy preserving technologies, in *1st International Conference on Wireless Communication, Vehicular Technology, Information Theory and Aerospace & Electronic Systems Technology, Wireless VITAE* (IEEE, 2009), pp. 336–340

73. S. Patel, D.R. Patel, A.P. Navik, Energy efficient integrated authentication and access control mechanisms for Internet of Things, in *International Conference on Internet of Things and Applications (IOTA)* (IEEE, 2016), pp. 304–309
74. P.P. Pereira, J. Eliasson, J. Delsing, An authentication and access control framework for CoAP-based Internet of Things, in *IECON 2014-40th Annual Conference of the Industrial Electronics Society* (IEEE, 2014), pp. 5293–5299
75. P. Pongle, G. Chavan, A survey: attacks on RPL and 6LoWPAN in IoT, in *International Conference on Pervasive Computing (ICPC)* (IEEE, 2015), pp. 1–6
76. S. Raza, S. Duquennoy, J. Höglund, U. Roedig, T. Voigt, Secure communication for the Internet of Things—a comparison of link-layer security and IPsec for 6LoWPAN. Secur. Commun. Netw. **7**(12), 2654–2668 (2014)
77. S. Raza, H. Shafagh, K. Hewage, R. Hummen, T. Voigt, Lithe: lightweight secure CoAP for the internet of things. IEEE Sens. J. **13**(10), 3711–3720 (2013)
78. F. Ren, J. Ma, Attribute-based access control mechanism for perceptive layer of the internet of things. Int. J. Digit. Content Technol. Appl. **5**(10), 396–403 (2011)
79. C. Rigney, S. Willens, A. Rubens, W. Simpson, Remote authentication dial in user service (RADIUS). No. RFC 2865 (2000)
80. R. Roman, C. Alcaraz, J. Lopez, N. Sklavos, Key management systems for sensor networks in the context of the Internet of Things. Comput. Electr. Eng. **37**(2), 147–159 (2011)
81. R. Roman, J. Lopez, C. Alcaraz, Do wireless sensor networks need to be completely integrated into the internet? in *3rd CompanionAble Workshop-Future Internet of People, Things and Services (IoPTS) Eco-Systems* (2009)
82. R. Roman, J. Lopez, S. Gritzalis, Situation awareness mechanisms for wireless sensor networks. IEEE Commun. Mag. **46**(4), 102–107 (2008)
83. M. Sain, Y.J. Kang, H.J. Lee, Survey on security in Internet of Things: state of the art and challenges, in *19th International Conference on Advanced Communication Technology (ICACT)* (IEEE, 2017), pp. 699–704
84. N. Saleh Al Marzouqi, ITU-T Study Group 20: IoT and its applications including smart cities and communities. Presentation, Hammamet, Tunisia (2016)
85. M.G. Samaila, M. Neto, D.A. Fernandes, M.M. Freire, P.R. Inácio, Security challenges of the Internet of Things, *Beyond the Internet of Things* (Springer, Cham, 2017), pp. 53–82
86. Z. Shelby, K. Hartke, C. Bormann, The constrained application protocol (CoAP) (2014)
87. S. Sridhar, A. Hahn, M. Govindarasu, Cyber-physical system security for the electric power grid. Proc. IEEE **100**(1), 210–224 (2012)
88. J. Srinivas, S. Mukhopadhyay, D. Mishra, Secure and efficient user authentication scheme for multi-gateway wireless sensor networks. Ad Hoc Netw. **54**, 147–169 (2017)
89. Study Group 11 at a glance. https://www.itu.int/en/ITU-T/about/groups/Pages/sg11.aspx. Accessed 8 Feb 2018
90. Study Group 13 at a glance. https://www.itu.int/en/ITU-T/about/groups/Pages/sg13.aspx. Accessed 8 Feb 2018
91. Study Group 16 at a glance. https://www.itu.int/en/ITU-T/about/groups/Pages/sg16.aspx. Accessed 8 Feb 2018
92. Study Group 17 at a glance. https://www.itu.int/en/ITU-T/about/groups/Pages/sg17.aspx. Accessed 8 Feb 2018
93. Study Group 2 at a glance. https://www.itu.int/en/ITU-T/about/groups/Pages/sg02.aspx. Accessed 8 Feb 2018
94. Y. Sun, H. Li, Efficient signcryption between TPKC and IDPKC and its multi-receiver construction. Sci. China Inf. Sci. **53**(3), 557–566 (2010)
95. R.T. Tiburski, L.A. Amaral, E. De Matos, F. Hessel, The importance of a standard security architecture for SOA-based IoT middleware. IEEE Commun. Mag. **53**(12), 20–26 (2015)
96. K. Toumi, M. Ayari, L.A. Saidane, M. Bouet, G. Pujolle, HAT: HIP address translation protocol for hybrid RFID/IP internet of things communication, in *International Conference on Communication in Wireless Environments and Ubiquitous Systems: New Challenges (ICWUS)* (IEEE, 2010), pp. 1–7

97. J.F. Valenzuela-Valdes, M.A. Lopez, P. Padilla, J.L. Padilla, J. Minguillon, Human neuro-activity for securing body area networks: application of brain-computer interfaces to people-centric internet of things. IEEE Commun. Mag. **55**(2), 62–67 (2017)
98. X. Wang, X. Sun, H. Yang, S.A. Shah, An anonymity and authentication mechanism for internet of things. J. Converg. Inf. Technol. **6**(3), 98–105 (2011)
99. W. Xie, Y. Tang, S. Chen, Y. Zhang, Y. Gao, Security of web of things: a survey (short paper), in *International Workshop on Security* (Springer, Cham, 2016), pp. 61–70
100. T. Yan, Q. Wen, Building the Internet of Things using a mobile RFID security protocol based on information technology, *Advances in Computer Science, Intelligent System and Environment* (Springer, Berlin, 2011), pp. 143–149
101. Z. Yan, P. Zhang, A.V. Vasilakos, A survey on trust management for Internet of Things. J. Netw. Comput. Appl. **42**, 120–134 (2014)
102. Y. Yang, L. Wu, G. Yin, L. Li, H. Zhao, A survey on security and privacy issues in internet-of-things. IEEE Internet Things J. **4**(5), 1250–1258 (2017)
103. X. Yao, X. Han, X. Du, X. Zhou, A lightweight multicast authentication mechanism for small scale IoT applications. IEEE Sens. J. **13**(10), 3693–3701 (2013)
104. N. Ye, Y. Zhu, R.C. Wang, R. Malekian, L. Qiao-min, An efficient authentication and access control scheme for perception layer of internet of things. Appl. Math. Inf. Sci. **8**(4), 1617–1624 (2014)
105. G. Zhao, X. Si, J. Wang, X. Long, T. Hu, A novel mutual authentication scheme for Internet of Things, in *Proceedings of 2011 International Conference on Modelling, Identification and Control (ICMIC)* (IEEE, 2011), pp. 563–566
106. K. Zhao, L. Ge, A survey on the internet of things security, in *9th International Conference on Computational Intelligence and Security (CIS)* (IEEE, 2013), pp. 663–667
107. K. Zhou, T. Liu, L. Liang, Security in cyber-physical systems: challenges and solutions. Int. J. Auton. Adapt. Commun. Syst. **10**(4), 391–408 (2017)
108. L. Zhou, H.C. Chao, Multimedia traffic security architecture for the internet of things. IEEE Netw. **25**(3), 35–40 (2011)
109. B.U.I. Khan, R.F. Olanrewaju, F. Anwar, R.N. Mir, ECM-GT: Design of efficient computational modelling based on game theoretical approach towards enhancing the security solutions in MANET. Int. J. Innov. Technol. Explor. Eng. (IJITEE) **8**(7), 506–519 (2019)
110. R.F. Olanrewaju, B.U.I. Khan, F. Anwar, R.N. Mir, M. Yaacob, T. Mehraj, in *Bayesian signaling game based efficient security model for MANETs*, ed. by K. Arai, R. Bhatia. Advances in Information and Communication. FICC 2019. Lecture Notes in Networks and Systems, vol 70 (Springer, Cham, 2019)

Critical Challenges in Access Management Schemes for Smartphones: An Appraisal

Tehseen Mehraj, Burhan Ul Islam Khan, Rashidah F. Olanrewaju, Farhat Anwar and Ahmad Zamani Bin Jusoh

Abstract A growing trend exerted by current users in accessing sensitive data and performing critical data exchanges predominantly highlights the proliferation usage of mobile phone devices by users for accessibility. There exists a demand for a security solution capable of thwarting the existing threats while offering extended support, at the same time conserving user adaptability. In this research, an intensive survey has been conducted in which various security solutions based on biometric and non-biometric access management schemes have been contemplated. A lack of absolute or standard access control management scheme capable of delivering a secure and feasible solution on mobile phones persists. Each of the works offered by researchers has been single-handedly evaluated. Finally, loopholes and open challenges were deduced from the study conducted.

Keywords Authentication · Access management · Biometric authentication · Security · Mobile device

1 Introduction

A growing trend is being observed among current users in accessing their crucial and sensitive data on smart mobile phone devices, which at the same time act as a critical platform for performing business transactions [1, 18, 26, 27, 32, 58, 70]. With time, smart mobile phone devices are delivering novel functionalities resulting in their wider adaptability and increased popularity among users while intensifying vulnerability to fraud [19, 42, 43, 54, 57]. According to research, 82% of people (age group: 25–35) together with 70% of domestic users perform their critical financial transactions via mobile phones [11]. There exists a user demand for increased security

T. Mehraj
Department of ECE, Islamic University of Science & Technology, Awantipora, Kashmir, India

B. U. I. Khan (✉) · R. F. Olanrewaju · F. Anwar · A. Z. B. Jusoh
Department of ECE, Kulliyyah of Engineering, International Islamic University Malaysia, Kuala Lumpur, Malaysia
e-mail: burhan.iium@gmail.com

M. Elhoseny and A. K. Singh (eds.), *Smart Network Inspired Paradigm and Approaches in IoT Applications*, https://doi.org/10.1007/978-981-13-8614-5_6

while preserving adequate user convenience. Such security solutions allowing secure access to sensitive personal information at the same time protecting banking related critical transactions should be able to thwart contemporary attacks while offering superior user adaptability [18, 50, 54, 64].

Security modernisation still employs the password-based or single-factor approach to achieve access management among numerous digital channels as in the event of compromise passwords offer easy revocation while presenting inexpensive deployment. The usage of user-sensitive passwords in the extensive study of unauthorised access to large-scale networking systems forms the paramount level of reliable and sensitive information [16, 21, 41, 75, 77]. In the current era, an individual has to maintain many digital identities for accessing services like for banks, emails, websites, shopping, social networking and so forth, each requiring passwords to be remembered by the users. Eventually, the individual users end up choosing simple passwords for easy memorization and in worst cases tend to use the same passwords for accessing all the services, resulting in increased vulnerability to fraud. Further, the availability of numerous password hacking tools, copious key-logger software and dictionary attacks have added to possibilities of violation [3, 24, 49, 51, 56, 73, 76]. Moreover, passwords can be easily forgotten, observed or shared, thus, forming an impractical security solution.

Numerous initiatives for replacing complex passwords have been presented in [2, 4, 30, 35], which offer dedicated, specially featured hardware tokens and smart cards, for achieving increased security but prove unmanageable, inconsistent while lacking in user ergonomics [52, 68]. Further, security solutions such as in [7, 47], fail to function on resource-constrained devices like mobile phones, which are increasingly employed for sensitive data transfer [27, 47]. Furthermore, a variety of facilities exist like cloud technology, which aim at offering many services to its customers. However, the data sharing approach utilised by the cloud technology uncovers many flaws and hence results in its susceptibility to numerous attacks [31, 48, 53, 55].

For digital vendors, user adaptability is a major concern. Customers demand simple user experience without compromising on the quality of services being offered. Higher intolerance is presented by younger generations towards solutions failing in their expectancies [6, 10]. In a report by the World Economic Forum (WEF), biometrics are expected to deliver a potential solution offering reliable security without compromising user conveniences in the financial sector [44].

Biometrics involves identifying individuals based on physical (e.g. fingerprint, face, iris, gait, etc.) or behavioural (e.g. user behavioural patterns, keystrokes) characteristics [28, 68]. In terms of user adaptability, there exists no comparison of security measures grounded on biometrics since biometric traits proffer advantages such as (i) Ease of use: the individual possesses the biometric feature giving him/her the freedom from remembering as in case of passwords or from carrying any token, (ii) Uniqueness: inimitability of biometric traits across individuals, (iii) Universality: every individual possesses the biometric characteristic, (iv) Permanence: biometric trait being unchangeable with time [34, 65]. However, with the upsurge of software and hardware resources on smart mobile devices regarding processing power, high definition cameras and fingerprint sensors, it is high time to incorporate these resources to implement security offered by biometrics on these devices.

Nevertheless, biometrics-based access control schemes can be breached owing to replay attacks in addition to spoofing attacks [5, 8, 9, 13, 29, 37, 59, 61, 62, 66, 68]. Biometric traits are not secretive; they are static in nature; one can take a picture of someone's face with or without their information and use it later to execute fraudulent access. Contrariwise, there exist solutions to safeguard against such threats in the form of Presentation Attack Detection (PAD) such as liveness detection, but they incur substantial computational cost hindering deployment on mobile phones [69].

According to NIST, widely used two-factor access management implementing SMS is no longer recommended [15, 33, 45]. Therefore, there exists a need for a system to be developed which provides the appropriate security on mobile devices without compromising user adaptability and at the same time has extended existence. To achieve this goal, a secure multimodal access management system combining biometrics as a primary credential with some non-biometric-based techniques should be developed.

2 Background Work

In this section, a detailed study of several common security solutions sectioned into biometric-based access management and non-biometric-based access management schemes offering security to mobile phones devices has been carried out.

2.1 Biometric-Based Access Management

Several access control mechanisms based on biometrics involving both physiological/physical and behavioural features of a person. Biometrics are being widely adopted owing to its uniqueness, ease of use and universality. The reviewed access control schemes centred on biometrics are as under.

2.1.1 Access Management Centred on Physiological Characteristics

The physical characteristics or features of an individual form the foundation for physiological based access management schemes. Fingerprint, retina/iris, face, palm geometry, form the physical characteristics which are comparatively unchanged with the ongoing time.

Face recognition
The access control schemes centred on face recognition involve exploitation of facial features obtained via digital image or video frames. In the field of mobile authentication, face recognition forms a potential area for research.

The authors [25] have presented an open and efficient Android-based face recognition scheme. The presented system incorporates face and eye detection by employing the following algorithms for feature extraction (LBP-Local Binary Pattern), preprocessing (ROI-Region of Interest), dimensionality reduction (LDA—Linear Discriminant Analysis and PCA—Principal Component Analysis) followed by minimum distance classifier (Euclidean distance). The entire system is primarily implemented on OpenCV (Open-Source Computer Vision) SDK, which provides efficient results as compared to using JavaCV together with Android SDK. ROI pre-processing step has been involved in achieving better accuracy. The experimental results revealed that LDA offered 96.0% accuracy while PCA delivers 93.8% accuracy in recognition. However, a small dataset has been considered, and the need for comparison with other state-of-the-art algorithms persists.

The authors [18] have proposed an offline biometric identification system. The offline face recognition has been implemented on a mobile device, and large datasets have been employed. OpenCV has mainly been utilised to realise the system on Android devices along with the usage of LBP face detection for feature extraction. Haar cascade classifier has been employed for performing pre-processing to increase the recognition rate and hence the accuracy of the system. Chi-square distance has been used to conduct face classification. The accuracy and the performance of the proposed system have been provided regarding matching error rates and response time. Identification accuracy with a probability of 87% is achieved in large datasets together with the response time of 30 s to search the biometric test sample among 500 biometric templates. However, a comparison with various available state-of-the art algorithms is desired as only LBP face recognition algorithm has been implemented and tested.

Replay attacks form a prominent threat to static passwords, however, in case of the biometric authentication system, one such research was conducted by Smith et al. [68], where a simple biometric replay attack has been implemented against current state-of-the-art face recognition systems. A method was proposed by the authors to counter replay attack aiming at face recognition on smart devices incorporated for commercial purpose by employing non-invasive challenge and response method. Conventional approaches address replay attacks by utilising distinctive sensors with dedicated processors and by functioning in controlled environments. However, the authors have considered systems operating in uncontrollable environments without any specialised dedicated security hardware. The vulnerability of the consumer devices based on biometric systems has been exposed as the authors have successfully conducted replay attacks, which effectively bypassed the liveness testing stage of the face recognition system. Face reflections are used to implement the challenge-response mechanism to counter-reply attacks. Face reflections are created in real time on the screen using colours reflecting the user's face. The system has been evaluated under ideal conditions, but more accurate image reflection methods need to be developed to deliver better security in real world situations. Various factors such as video image resolution, screen illumination levels, environmental conditions and camera settings are not considered.

On smart mobile phone devices, face in comparison to iris and fingerprint evolves as a prominent biometric trait to be contemplated [19, 36]. Although access control schemes centred on face recognition are widely accepted by the current smartphone users, yet there exist certain areas, which require further improvements in terms of improved accuracy and performance in realistic uncontrollable situations. One such solution will be consideration of extensive datasets, offering an excellent training set that covers the types of variations that one expects to occur in the testing set, hence yielding better results.

Fingerprint recognition
Access control on smartphones via fingerprint has turned out to be a potential research area. The security schemes centred on fingerprints have been previously applied on smartphones and proved to be a user–adaptable solution for verification of a person's identity.

A secure, robust and low-cost biometric authentication system on mobile smartphones has been presented by Rathi and Sawarkar [63]. The system makes use of fingerprint, which is a unique feature, to provide access to services, thus freeing users from the cumbersome task of remembering passwords. The authentication mechanism aims at offering secure access to legitimate users within a short frame of time to prohibit attacks from intruders. Instead of using additional hardware in the form of sensors, which increase the complexity of the system and to eliminate the extra cost, the scheme makes use of inbuilt cameras present in mobile phones. The scheme has been developed utilising Android and OpenCV (Computer Vision) library followed by an RGB matching algorithm. Illumination variation effects are taken into consideration and are reduced by maximising the colour balance. However, the system lacks in obtaining crucial feature as low definition camera has been employed. Further, the camera used for acquiring fingerprint cannot be considered a stable fingerprint acquiring sensor as the user may have to take into consideration some factors like background, lighting, orientation etc. which otherwise may reduce the accuracy of matching algorithm but at the same time hinder user adoptability. Higher false acceptance rate and 77–78% accuracy is possible initially with the false acceptance rate of 33% but smaller dataset has been considered, and performance analysis regarding the accuracy of the system and the resources consumed on mobile devices has not been performed.

The authors [17] have proposed an Android-based biometric authentication scheme for mobile phone devices. The main aim of the authentication scheme is to compare the two images of fingerprints, one that is provided by the user and the other that is registered in the database. For fingerprint, processing three authenticating algorithms have been presented while being evaluated as per their speeds and accuracy rates. The first algorithm considers singularity points extraction using Poincare indexes. The second algorithm exploits minutiae extraction followed by a sturdy matching algorithm, and the third one employing minutiae extraction followed by a fast matching algorithm. The respective three algorithms have been evaluated regarding user adaptability, acceptable identification tolerance, speed and system usability. However, the sensor used for fingerprint acquisition on mobile devices lacks in the complete recovery of the fingerprint thereby reducing the recognition rate and hence user adaptability.

Consequently, the access control schemes centred on fingerprint-based approaches tend to offer a lightweight and economical solution for resource-constrained mobile phones. However, the complete acquisition of the entire fingerprint is not possible from the offered hardware in present-day mobile phone devices in addition to the incompatibility of the state-of-art algorithms in the presence of cut marks or dirt. Thus, making fingerprint biometrics as a weak contestant in the formation of an adaptable and secure solution.

Heart Sonic Waves

An exposition has been provided for systems requiring continuous authentication by Andreeva [7]. In continuous authentication, user accesses are controlled at regular intervals of time. In this paper, the solution is based on Body Area Networks (BAN) designed primarily for information security systems associated with various life-supporting medical devices. In other words, the system delivers continuous authentication for life-support medical devices. The authentication method used employs human heart sonic signal waves as a criterion for authentication. Using heart sonic signals provides a higher acceptability degree in medical devices as it is independent of the user's action. The reliable authentication system is provided by taking precise data from heart sonic waves exploiting BAN embroiling benefit of heart sonic signals, which are difficult to be lost and forged in the absence of specialised types of equipment.

Further, the distinctive characteristics of heartbeat sound are extracted employing standard speech recognition algorithm taking into consideration the specificity of the heart sound. However, prolonged authentication cannot be expected by the system as the heart sound conditions of humans do not remain the same with time, owing to the psychological conditions of a person. Further, the proposed authentication system is primarily established on the BAN, which involves complex construction and various sensors, thus hindering adequate user adoptability. In addition to this, the risk of intrusion exists during data transmission, which eventually cuts back the efficiency of the system.

Further, these physiological authentication security schemes besides offering adoptable user solutions must also deliver strong resistance towards replay attacks. Hence, one such solution in the form of liveness detection appears for such security schemes. The authors [23] offer a novel software providing a fake detection technique that can be employed in numerous biometric methods to identify diverse forms of fraudulent access. Liveness assessment has been used in a user-friendly, fast and non-intrusive way by utilising image quality assessment. The proposed scheme provides a low complexity degree making it suitable for mobile phones by considering the 25-general image quality feature. The system has been evaluated against

Table 1 Performance comparison of physiological biometric authentication techniques

Author	Technique	Performance		
		FAR	FRR	Accuracy (%)
Jiawei et al. [25]	Face recognition	–	–	93, 96
Darwaish et al. [18]		–	–	87
Kavita et al. [63]	Fingerprint recognition	33%	–	77
Conti et al. [17]		–	–	93
Andereeva et al. (2012)	Heart sonic waves	–	–	–

various datasets of the face, iris and fingerprints and provides better results than the state-of-art methods in identifying fake traits.

Further, the system presents a very low degree of complexity, which makes it suitable for real-time applications. Implementation is performed on computer systems and not on mobile devices. Also, the state-of-the-art algorithms of various biometric traits have not been specified.

A substantial and significant advantage offered by physiological biometric authentication schemes is the uniqueness of the features, which forms a stable base to develop reliable authentication solutions. Each physiological biometric-based authentication scheme discussed above has its own benefits and shortcomings. However, the feasibility of such schemes on mobile phone devices needs some light.

In Table 1, results obtained from the related studies indicating authentication accuracy have been outlined. It is evident that the face recognition-based authentication schemes can attain high accuracy. Nonetheless, some common issues exist, which need to be taken care of while deliberating face recognition on mobile phone devices. One such concern comprises liveness detection, which forms an essential factor in increasing mobile phone devices' implementation cost. Another matter to be contemplated is that such systems do not support active authentication, leaving a chance for intruders to compromise the security on mobile devices.

2.1.2 Access Management Centred on Behavioural Characteristics

Behavioural traits such as behaviour profiling, gait, voice, signature and typing rhythm or keystroke dynamics form the basis for access management schemes founded on behavioural biometrics.

Behaviour Profiling

Behaviour profiling involves the interaction or behaviour of an individual to their respective mobile devices while availing numerous services; for instance, location, application usage, text message and voice call information. Such methods make use of machine learning technology by comparing users' current activities with the profile based on historical usage. Transparent active authentication can be easily achieved by such systems.

A robust verification process has been proposed by Fridman et al. [22], which handles the human-device interaction which is dynamic in nature. Four parameters namely: (i) location (Wi-Fi-indoors, GPS-outdoors), (ii) text entered, (iii) apps usage, (iv) websites visited, have been contemplated. The paper has proposed a decision fusion, which asynchronously integrates the four parameters and provides decisions based on serial authentication. In this paper, the proposed multimodal continuous authentication system is evaluated by specific characteristics, which include error rates of judgments rendered by local classifiers together with the global decisions and the influence to the fused decision by every local classifier. Chair-Varshney optimal fusion ruling has been employed for the combination of offered multimodal decisions. Location-based classifier utilises Support Vector Machine (SVM) with kernel function as the radical basis function. The proposed system monitors well the energy constraints as the modalities chosen result in low power consumption, hence resulting in better performance than the conventional authentication systems. The decision support system utilised by the proposed method is flexible enough to strengthen the decision level approach without having to change the underlying fusion directive. The dataset employed includes the vast pool of subjects cogitating around 200 for five months along with the portfolios of modalities, which were concurrently tracked with a synchronised timestamp. The temporal performance of the intruder detection has been characterised unlike in existing systems. The proposed authentication system has been able to achieve 0.05 (5%) of Equal Error Rate (ERR) within 1-min user interaction and 0.01 (1%) of ERR after 30-min user interaction. The contribution to fused global decision and performance of each local classifier has been presented. The location-based classifier has been identified as the most dominant factor in the fusion systems' performance. However, since the system utilises four modalities out of which location-based pattern is the highest contributor as the location classifier has used GPS, precise authentication cannot be achieved as there exist some services within the same geographical location, which require authentication. For instance, within the same building, there are some services, which need authentication, but the user may be allowed to access.

Two practical problems related to Implicit Authentication (IA) have been addressed by Yang et al. [73]. The system does not require the explicit user actions. Further, the method aims to enhance the user experience to protect devices from more misuse. The authentication mechanism utilises behavioural data of numerous types to deduce user behaviour. The author seeks to solve the two crucial issues associated with the implicit authentication. The first problem is being able to achieve a mechanism for obtaining the best possible retraining frequency while the user behaviour model is being updated. To solve this problem, an algorithm utilising Jensen–Shannon (JS) distance is proposed to deduce optimum retraining frequency. The second problem involves the ability to handle false negatives, i.e. when the legitimate user is not authenticated. For this, a dynamic privilege scheme also based on JS-distance allowing multilevel fine-grained access mechanism has been introduced. The method proposed has been found to be able to automatically select the best retraining frequency for each who is varying in nature and further handles the false negatives better than the conventional systems by locking only part of the system rather than locking the whole system, thus, adding to the better user experience.

Moreover, depending on the current behaviour of the user, the level of privilege assigned to him/her is decided rather than the predefined privilege rule as followed by the conventional systems plus the consequences of failure resulting from false negative authentication have been successfully decreased. However, the proposed scheme has not shown the performance on smartphones on which the implicit authentication is widely implemented. Also, the dataset has considered nine types of data involving accelerometer, GPS, app installation, SMS, call logs, battery usage, Bluetooth devices, log and applications running. Better data should be considered to strengthen the implicit authentication systems further so that IA can be used in services that require strong security. Vulnerability to mimic attacks, synthetic attacks, sensor-sniffing attacks, smudge attacks, accelerometer side-channel attacks, timing attacks, keystroke inference attack, location inference attacks and gyroscopic side-channel attacks persist.

A novel access control scheme founded on specific user context has been developed by Shebaro et al. [67], capable of dynamically revoking and granting privileges to device users. The authors have enforced and specified the context-based access control mechanism by modifying the Android operating system. Context-Based Access Control (CBAC) policies are implemented on the Android operating system and users are provided with the tool that permits users to outline various physical places by capturing Wi-Fi parameters. In other words, the configuration policies are set by the user of the smartphone over its application's usage of device services and resources concerning different contexts. Sensitive data leakage and privacy breaches resulting from providing access to malicious apps have been restricted to existing related works which aim merely at detecting overprivileged applications. Thus, they fail to recognise buggy or malicious applications.

Further, the CBAC has been able to achieve the location information more accurately as compared to conventional approaches. In this paper, the CBAC policies have been implemented in mobile phone devices bearing in mind the battery usage and performance overhead. CBAC system has performed security analysis in which it has been successfully able to mitigate the threats. However, 16% false positives, which imply unregistered areas appearing as user-defined have been found, that need to be further reduced. Moreover, the system suffers from memory overhead, which mainly occurs because the location module is being continuously instantiated within predefined intervals to keep the context of mobile up-to-date.

Such schemes tend to offer a feasible solution for smartphones; however, such systems tend to perform abnormally when the user interacts with unexpected behaviour than usual.

Keystroke Dynamics

In this approach, a person's typing speed, manner or rhythm are being exploited. An enhancement to the keystroke dynamics access control schemes on smartphones has been achieved by Tsai et al. [72] while improving user ergonomics. In other words, the burden on users to again receive the training of keystroke dynamics whenever they wish to change their passwords is eradicated. The system can predict the possible keystroke dynamics of authentic users based on the limited data determined during the training process thus enabling a user to change their PIN at any moment without the necessity to retain. However, the system discards the retraining phase which otherwise plays a crucial role in maintaining the accuracy of machine learning as time evolves. Authentication failure is not being handled appropriately, that is, when a legitimate user has rejected authentication; it leads to reduced system efficiency. Further, susceptibility to mimic attacks, synthetic attacks, sensor-sniffing attacks, smudge attacks, accelerometer side-channel attacks, timing attacks, keystroke inference attacks, location inference attacks, and gyroscopic side-channel attacks etc. still perseveres.

In comparison to physiological based access control schemes, behavioural biometrics without any additional hardware requirements tend to offer a solution to achieve active authentication transparently, thus, providing with a less costly solution than physiological authentication. However, as seen in Table 2, the accuracy of such systems has been found to deliver inconsistent results when a user acts inversely than his/her natural behaviour.

Physiological based access control schemes emerge susceptible to replay attacks. Besides, there exists a need for enhancing security in access control schemes centred on behavioural traits. Combining two or more of these features will strengthen security while thwarting such vulnerabilities. Merging biometric with non-biometric or physiological with behavioural can be done to improve the complete system.

Several studies done in this context includes [47], where the authors have proposed a novel authentication framework implementing multimodal biometric user authentication. The existing authentication mechanism based on biometric schemes used especially in touchscreen smartphones has been complemented along with introducing an innovative framework employing both the biometric (touchscreen dynamics) and non-biometric (PIN) techniques, thus enhancing both security and usability than the conventional related systems. The authors have successfully carried out the experiments, which support the fact that multimodal biometric approach can be employed to lower the misleading rates of authentication systems based on a single biometric

Table 2 Performance comparison of existing behavioural biometric authentication techniques

Author	Technique	Performance		
		FAR (%)	FRR	ERR (%)
Tsai et al. [72]	Keystroke dynamics	–	–	0.18
Shebaro et al. [67]	Behavioural profiling	16	29%	–
Yang et al. [73]		64	–	–
Fridman et al. [22]		–	–	5
Javier et al. (2015)		–	–	20.73
Tsai et al. [72]	Keystroke dynamics	–	–	0.18

system on touch-enabled phones. However, the two significant drawbacks associated with biometric authentication are the speed and accuracy; the touch screen dynamics considered in the framework demand high performance and accuracy, which is difficult to achieve on mobile devices. Also, the system emerges to be vulnerable to touch-logger detection attacks.

A multimodal scheme combining physiological fingerprint with a behavioural fingerprint has been presented by Teo and Neo [71]. Transparent and continuous authentication is provided during the entire process even if the fingerprint is lost, as the system transforms the static and unique fingerprint physical feature into the behavioural characteristic, which has qualities of being irrevocable and exclusive at the same time. However, the proposed system manipulates human knowledge with individual fingers, which affect user ergonomics as the sequence needs to be memorised by the user. Further, the first sensor for fingerprint acquisition is desired which comes with additional hardware requirements.

A multimodal biometric scheme has been offered by Raja et al. [60], employing periocular, face and iris for verification. The proposed system has been evaluated and analysed using both the multimodal as well as unimodal approaches. The presented method explores numerous score level fusion schemes to utilise the complementary information specified from the three modalities. The system has considered different scenarios including varying illumination conditions while considering face recognition and exploiting other features such as iris and periocular to provide better recognition scores. Further, the authors have successfully implemented Open Source Iris Segmentation Algorithm (OSIRIS v4.1) on the Android platform. An extensive dataset of 78 subjects has been considered plus 0.68% Equal Error Rate (EER) has been achieved from the experimentations, certifying the robust functioning of the offered system. Although, the face has been the principal contributor in the proposed scheme, different distances to a camera, camera properties, face angles and expressions have not been considered, which will eventually affect the remaining modalities like iris and periocular. Iris has been regarded as one of the patterns, which demand high definition cameras along with extremely controlled conditions for acquisition and incurs high computation cost. Also, the performance analysis regarding the limited resources of the smartphone such as memory and battery consumption has not been undertaken.

Novel biometric multimodal security mechanism has been introduced on mobile devices which offer identity and verification management for different security levels [19]. An Android application centred on multimodal biometric recognition of iris and face has been designed to be embedded on mobile devices. Varying lighting conditions and poses are handled by the selection of the best biometric sample employing iterative entropy evaluation approach. Apart from the usual steps in the acquisition of face and iris, an additional action has been considered called the anti-spoofing step, which follows the segmentation stage to safeguard numerous security-critical applications. The anti-spoofing stage performs the liveness detection by exploiting 3D-geometric collinearity invariants, which calculates the ratio of determinant matrices. Linear Discriminant Analysis (LDA) has been used for face feature extraction as well as matching.

Further, the optimisation of algorithms has been done to form a computationally light and low demanding solution for mobile devices, which have limited resources. However, high-resolution cameras are needed for iris detection along with considering different camera properties followed by extremely controlled situations that hinder adequate user adoption. Equally, the results obtained by this fusion model do not offer convincing results. Table 3 further outlines the merits and weaknesses of every biometric technique.

On the whole from the survey conducted, it has been found higher accuracy is offered by physiological authentication techniques than the behavioural authentication techniques resulting in better performance while at the same time being widely accepted by users. However, hardware requirements in the case of individual physiological features such as iris are desired. In contrast, behavioural biometrics offer a solution to attain transparent active authentication without any additional hardware requirements forming a cheaper solution. Combination of the two can be viewed as a promising solution to overall improve the performance of biometric authentication systems.

The tabulation of security solutions offered by numerous researchers is below to benchmark. Table 4 highlights the various contributions followed by associated limitations as follows.

Table 3 Pros and cons of various biometric techniques

Biometric techniques	Pros	Cons
Face recognition	• Ease of use and high user acceptability • Non-intrusive in nature with relatively good accuracy • Additional hardware not needed • Ease of use	• Poor performance regarding varying lighting condition, face angle etc. • Additional hardware required
Fingerprint recognition	• Low battery power consumption on resource-constrained mobile phone devices	• Poor performance in the presence of dirt and cuts on a finger • Provides inconsistent accuracy
Keystroke dynamics	• Additional hardware not needed • Offers continuous authentication	• Deteriorates system performance during unusual user behaviour
Behaviour profiling	• Delivers continuous authentication	• Deteriorates system performance during unusual user behaviour
Heart sonic waves	• Provides continuous authentication	• Expensive implementation • Lacks in user ergonomics

Table 4 Review of biometric-based systems

Author	Contribution	Findings	Limitations
Andree [7]	Designed a continuous access management scheme utilising heartbeat sounds	• Presented a continuous access control for medical life-support devices • Heart sonic waves are difficult to be forged, hence constitute a better biometric trait • Precise information is retrieved from heart sonic waves using BAN to avoid any variation	• Persistent access management can't be achieved due to changing heart conditions with ageing of an individual • Involves complex implementation, hindering adoptability • Data transmission not secure
Kavita et al. [63]	Presented an affordable and robust access management system on mobile phones	• Considered environmental factors like illumination variations while using colour balance to reduce the same	• Incomplete acquisition of vital features due to use of inadequate hardware • Factors encountered in real world situations like lighting, dynamic background and orientation are not considered. Thus, hampering its deployment • High FAR-False Acceptance Rate and performance evaluation regarding resource consumption on mobile devices not done • Considered smaller data-set
Conti et al. [17]	Offered Android-based biometric access control scheme	• Investigated algorithms in terms of user adoptability, system usability and verification tolerance speed	• Incomplete acquisition fingerprint resulting in reduced accuracy

(continued)

Table 4 (continued)

Author	Contribution	Findings	Limitations
Darwaish et al. [18]	Designed offline face recognition on mobile phones	• Eliminates issues associated with online biometric schemes • Real-time face detection by implementing LBP	• Evaluation with numerous existing state-of-art systems is anticipated
Tsai et al. [72]	Designed an enhanced keystroke access management scheme	• Enhances user adoptability	• Eliminates retraining phase • Verification failure not handled properly • Vulnerable to numerous attacks like synthetic attack, accelerometer side-channel attacks, mimic attacks, smudge attacks, gyroscopic side-channel attack, keystroke inference attack and location inference attacks
Daniel et al. [68]	Designed a technique for face recognition offering protection against replay attacks	• Thwarted replay attack aiming face recognition on smart devices using non-invasive challenge and response method • Employed face reflections to implement the challenge-response mechanism • A recent state-of-the-art face recognition system has been considered	• Factors encountered in real-world situations like lighting, dynamic background and orientation are not considered. Thus, hampering its deployment • Implementation done on computer systems and not on smartphones

(continued)

Table 4 (continued)

Author	Contribution	Findings	Limitations
Fridman et al. [22]	Introduced an active access management scheme on mobile devices	• Provides robust verification process handling the dynamic human-device interaction • Utilised decision support system which is flexible enough to strengthen the decision level approach without the need to change the basic fusion directive • The contemplated dataset including the large pool of subjects has been considered, around 200 for a period of 5 months along with the portfolios of modalities which were concurrently tracked with a synchronised timestamp • Temporal performance of the intruder detection has been characterised • Achieved 0.05 (5%) of Equal Error Rate (ERR) within 1-min user interaction and 0.01(1%) of ERR after 30-min user interaction	• GPS acts as location classifier, failing to present precise location with respect to access management
Javier et al. (2015)	Designed a fake detection system capable of thwarting fraudulent access to several biometric schemes	• Employed liveness assessment in a user-friendly, fast and non-intrusive way by image quality assessment • Low complexity degree making it suitable for mobile phones by considering the 25-general image quality feature	• Implementation done on computer systems and not on smartphones • The considered state-of-the-art algorithms not specified

(continued)

Table 4 (continued)

Author	Contribution	Findings	Limitations
Jiawei et al. [25]	Designed an Android-based face recognition technique entitled XFace	• Employed LBP for face detection along with Fisherfaces and Eigenfaces being utilised and tested for face recognition • A region of Interest (ROI) pre-processing step involved to achieve better accuracy	• Evaluation with numerous existing state-of-art systems is anticipated • Contemplated small dataset
Kiran et al. [60]	Presented a multimodal biometric access management scheme employing face, periocular and iris characteristics	• Considered an extensive dataset of 78 subjects • 0.68% Equal Error Rate (EER) has been achieved from the experimentations	• Factors encountered in real-world situations like lighting, dynamic background and orientation are not considered • Iris acquisition incurs high definition camera with increased processing hampering its deployment on smartphones • Performance evaluation regarding resource consumption on mobile devices not done
Maria et al. [19]	Designed Android-based multimodal biometric security scheme	• An Android application centred on multimodal biometric recognition of iris and face has been designed on mobile devices • Safeguards numerous security-critical applications by incorporating additional anti-spoofing step	• Iris acquisition incurs high definition camera with increased processing with restriction of surrounding settings, thus hampering its deployment on smartphones
Meng et al. [47]	Presented multimodal biometric access control scheme	• Enhanced both security and usability by incorporating the non-biometric and biometric techniques forming a multimodal authentication system	• High accuracy and performance are desired for incorporating touch screen dynamics, which is hard to attain in mobile phones

(continued)

Table 4 (continued)

Author	Contribution	Findings	Limitations
Shebaro et al. [67]	Designed a dynamic access control scheme capable of assigning or revoking privileges to users dynamically	• Sensitive data leakage and privacy breaches resulting from providing access to malicious applications have been restricted • Capable of achieving the location information more accurately as compared to conventional approaches • Implementation of the respective CBAC policies on mobile phone devices regarding battery usage and performance overhead • Effectively mitigated the threats as shown by security analysis	• Vulnerable to touch-logger detection assaults • Experiences memory overhead
Yang et al. [73]	Focused on addressing two prominent issues with implicit authentication schemes	• Better user experience • The current behaviour of the user decides the level of privilege being assigned to the user • Successfully being able to cut back the failure resulting from false negative authentication	• Performance evaluation on mobile devices not presented • Incorporates data-set with weaker modalities • Vulnerable to numerous attacks like synthetic attack, accelerometer side-channel attacks, mimic attacks, smudge attacks, gyroscopic side-channel attack, keystroke inference attack and location inference attacks
Teo et al. [71]	Presented a framework combing physical fingerprint with behavioural fingerprint	• Transparent and continuous authentication is provided during the entire process even if the fingerprint is lost	• Reduced user ergonomics • Added hardware requirement in the form of primary sensor for fingerprint acquisition

2.2 Non-biometric-Based Access Management

Apart from biometric-based access management schemes, a large amount of research has been conducted on schemes centred on social networks, modified passwords, location information, smart devices, permission control and public key cryptography on mobile devices.

The authors [14] tend to strengthen the password-based access control schemes. Since users have to manage some different accounts, which include multiple email, accounts, bank accounts etc., there exists a high probability of compromising the security; as if one password is breached possibility exists there that all of the passwords will be violated. Furthermore, users prefer simple passwords so that they are easy to recall which makes them vulnerable to various attacks. A security mechanism has been proposed by the authors that aim to maintain the delicate balance between security and user-friendliness. Random free text has been added to increase the difficulty level in breaching the password while protecting the system from a pre-computed dictionary as well as rainbow attacks. However, the system suffers from authentication failure that result from the cached passwords in various newest browsers and other software. User adoptability is not up to the mark, as more things need to be remembered by users, unlike the existing system. Moreover, there exists vulnerability to keystroke attack.

The authors [74] attempt to enhance access control mechanism in LAN by exploiting USB and a UAT (User Access Control Table). This approach improves LAN's access control by making use of UAT and USB. Public key cryptography has been utilised to deliver identity authentication. USB provided to the user contains the private key and the digital certificates to private key, which is in a non-readable format to prevent an attacker who can steal private key from the user. The security of the authentication system is ensured by taking a random number comprising of the same digits as the hash value of corresponding information instead of exploiting authentication information and the resulting hash value. In the conventional approach, ciphertext was fixed which the attacker could capture and after that perform replay attack hence cracking the legitimate user's password. The proposed system prevents replay attacks by storing the authentication information on the USB itself. While Public Key Infrastructure has been employed which incurs higher cost, the system has been able to achieve the balance between performance and security by utilising additional hardware in the form of USB. However, this added hardware required to use the facilities of the proposed system results in degrading user adaptability.

In the study conducted by Durmus and Langendoen [20], a Wi-Fi access control scheme has been proposed utilising social networks. In other words, the proposed system automates Wi-Fi authentication by allowing people to share their passwords for their Wi-Fi access points without requiring manual involvement. Certificates are used to authenticate the user by the access point itself by verifying the signature of the certification authority. A scalable solution has been provided by limiting the search domain to devices nearby and friends on the social network. The search space is sorted and tied to the time of arrival and the presence of direct friends. The proposed system is based on the EAP-TLS and Web-ID connecting devices to distributed social networks, which make use of Wi-Fi probe requests to define the collection of neighbouring devices. The search complexity of social networks for indirect friends being quadratic in nature is the primary challenge in the proposed system. However, the social network search complexity has been decreased by using a collection of direct friends to bridge the access point to indirect friends exploiting the context information. The proposed decentralised approach can be employed with any other authentication protocol that encompasses certificates for instance SSH, Wi-Fi direct and IPsec. The proposed system requires that each device must have a profile on the social network, which represents its owner. However, it is not necessary that every person has a profile on the web.

Moreover, privacy concerns may be raised as complete friendship information is anticipated to be open for all. Furthermore, security concerns may arise because of worn out caches, which are used to perform offline authentication. Likewise, ownership related issues are not contemplated. The access control can be automated by applying Web-ID protocol to any certificate-based authentication. However, to decrease the complexity associated with Distributed Online Social Network (DOSN) search for trust, context-aware solutions are crucial to being designed.

A secure and effective access control scheme established on smart cards has been presented by Aboud [2]. The scheme allows enhanced user adoptability by allowing them to modify passwords offline. At the same time, the scheme thwarts server attack by shifting user authentication to the registration centre, which ensures the server acquiring the secret key, consequently delivering a practical and more secure solution. Also, the system performs the comparative analysis with relative systems which is found to be more cost-efficient.

The authors [39], presented a bilateral recurring access control scheme, i.e. ZEBRA (Zero Effort Bilateral Recurring Authentication). A bracelet as a hardware device has been utilised for providing continued authentication. The bracelet being worn by the user on its wrist is used to send signal continuously to the terminal device to confirm user presence. However, additional hardware requirement exists in this system. The accuracy of 85% has been achieved in identifying the correct user under ideal conditions with an identification time of 11 s. Nevertheless, the system lacks behind from user adoptability perspective and identification time raises to 50 s when acquiring a balance between accuracy and user adoptability.

The authors [38] have manipulated location information for providing a user access control scheme. The system performs the user identification particularly on mobile devices by exploiting knowledge of historical location traces of the user concerning time. The scheme implements Hidden Markov Model complemented with the marginal smoothing technique for location authentication. The system delivers 20.73% of Equal Error Rate—EER.

Furthermore, the authors [12] have put forward a mechanism for preventing sensitive data leakage on Android smartphones due to malicious applications. The system implements J8-classification algorithm to detect the presence of malicious apps and has achieved an accuracy of 98.6%. However, the evaluation of a vast amount of data of the proposed system should be done. Table 5 has outlined the contribution and drawbacks of various non-biometric schemes considered formerly as follows.

Table 5 Review of non-biometric schemes

Author	Contribution	Findings	Limitations
Zhang and Pie [74]	Introduced a novel access control scheme enhancing LAN's access control founded on User Access Control Table (UAT) and USB in local area network (LAN)	• Thwarts replay attacks by storing the authentication information on the USB itself	• Additional hardware in the form of USB is required • Lacks in user ergonomics
Aboud et al. [2]	Designed a secure and effective authentication scheme based on smart cards	• Susceptible to a multitude of threats, e.g. offline dictionary attack, stolen attack, server attack, user attack, etc. • The system appears to provide more flexibility to the users since user anonymity is well taken care of • Password change doesn't involve dependence on registration centre • Cost-efficient as rivalled to current systems	• Expensive solution for vendors as it incorporates additional hardware: smart card
Durmus et al. [20]	Introduced access management scheme for Wi-Fi by exploiting social networks	• Abandons the centralised approach which entrenches social networks for Wi-Fi authentication hence eliminating single point failure problems and privacy concerns • The social network search complexity has been decreased • The proposed decentralised approach shows compatibility with other authentication protocols that encompass certificates such as SSH, Wi-Fi direct and IPsec	• The proposed system requires that each device must have a profile on the social network which represents its owner • Privacy concerns may be raised as complete friendship information open for everyone • Worn-out caches lead to security issues

(continued)

Table 5 (continued)

Author	Contribution	Findings	Limitations
Shrirang et al. [39]	Designed an active access management scheme i.e. ZEBRA	• Provides better solution in active authentication field by making precise authentication of the user • Achieved accuracy of 85% in identifying the correct user under ideal conditions with an identification time of 11 s	• The system lacks from user adoptability perspective • Additional hardware requirement in the form of a bracelet
Upal and Rama [38]	Offered an access management system utilising trace history	• Presented a novel algorithm as Hidden Markov Model accompanied using marginal smoothing method for location authentication • Outperforms other techniques regarding Equal Error Rate (EER) of 20.73%	• Vulnerable to spoofing and replay attacks
Chowdhury et al. [14]	Forwarded a technique supporting the access management schemes based on passwords	• Level of difficulty in breaching the password is increased by adding free random text • Usage of random text makes the system immune against pre-computed rainbow and dictionary attacks in addition to shoulder surfing and replay attacks	• Verification failure arises due to passwords cached in various newest software and web browsers • Reduced user adoptability • Vulnerable to keystroke attack
Yavuz et al. [12]	Introduced a mechanism hindering sensitive data leakage	• Utilised J8 classification algorithm • Achieved 98.6% accuracy in detection of benign and malicious applications	• Contemplated Smaller data-set

3 Research Challenges and Open Issues

Subsequent contemplation of numerous security solutions available in access management revealed numerous unattended loopholes and challenges ranging from computational complexity to security vulnerability to user adaptability. Finally, the progressive study will conclude with the ensuing open issues that need to be addressed (shown in Fig. 1):

- The security solutions [2, 7, 19, 39, 63, 71, 74] require additional dedicated hardware and smart cards, which results in user inconvenience, hence impeding their adaptability.
- Several concerns in the reviewed security schemes [7, 47] have been revealed such as elevated computational and processing time, thus making these schemes incompatible with mobile devices.
- Numerous schemes [12, 25, 60, 63] delivered execution on smart mobile devices although their performance regarding large datasets has not been indicated.
- The accuracy of systems [22, 67, 73] based on behaviour authentication largely decreases if a user behaves differently, hence damaging the performance of biometric behavioural authentication systems.
- The immense popularity of smart mobile devices makes them vulnerable to numerous attacks and threats such as touch-logger/key-logger, liveness detection, mimic attacks etc. Likewise, security schemes, particularly those based on behavioural authentication [14, 20, 47, 72, 73], fail to protect against such threats.

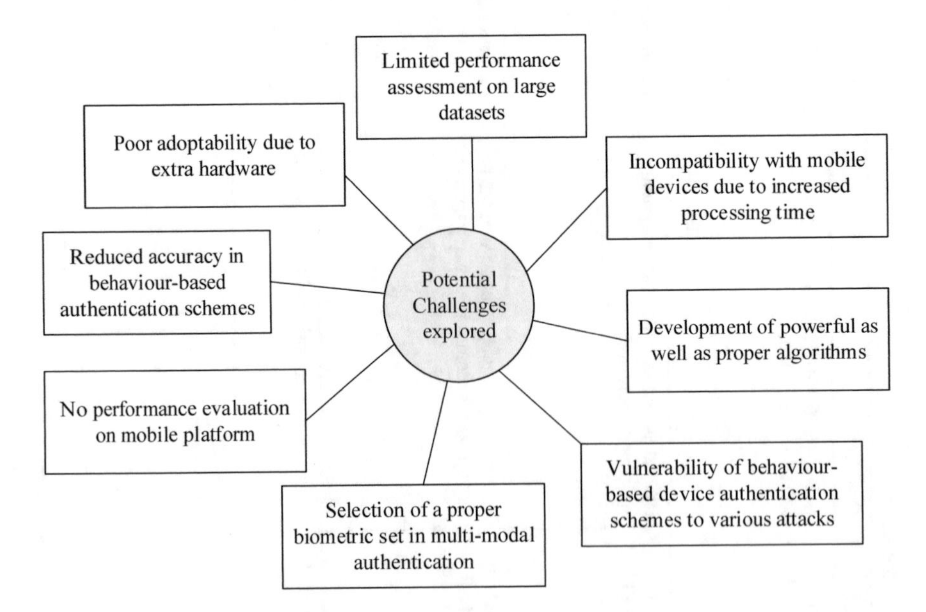

Fig. 1 Potential challenges explored in existing authentication schemes

- Designing reliable authentication systems capable of selecting or deciding an appropriate biometric set is an open problem particularly when considering access control schemes based on multimodal approach [12, 19, 22, 47, 60].
- Schemes like [22, 38, 73] undertake behaviour profiling where the performance heavily depends on algorithms designed to produce pattern classification models. Developing appropriate and powerful algorithms is again a major issue.
- Performance evaluation particularly on mobile platform, which is increasingly growing popularity for the exchange and storage of sensitive information, is not conducted by most of the studied authentication schemes [14, 20, 23, 60, 63, 67, 68, 73, 74]. Therefore, there is an immediate need to evaluate available security schemes on mobile devices.

4 Conclusion

An evaluation of former access control schemes focussed on non-biometric and biometric approaches have been presented in this paper. The significant impacts and weaknesses of the similar security schemes have been enumerated explicitly. All the reviewed access control schemes lacked in one setting or the other. The advancement of computing power on mobile devices and their popularity among people in performing critical and sensitive data exchange is uncovering an urgent need for a highly secure solution than the existing ones. Out of the numerous security solutions which have been analysed, those utilising the biometrics are the ones offering ameliorate user adoption, i.e. biometrics has been found to provide a superior range of access control techniques. By reviewing the security schemes that have been put forward in the past as well as the ones that are currently in use, it can be concluded that the ones with biometrics ensure a new secure method of accessibility for both, private as well as public networks. However, biometrics can be easily breached via reply attacks and hence are capricious. Therefore, the utilisation of the biometrics needs to be dealt with more attention to put forth systems with enhanced user adaptability and security.

Acknowledgements This work was partially supported by the Ministry of Higher Education Malaysia (Kementerian Pendidikan Tinggi) under Research Initiative Grant Scheme number: **RIGS16-334-0498**.

References

1. N. Abbas, Y. Zhang, A. Taherkordi, T. Skeie, Mobile edge computing: a survey. IEEE Internet Things J. **5**(1), 450–465 (2018)
2. S.J. Aboud, Secure password authentication system using smart card. Int. J. Emerg. Trends Technol. Comput. Sci. (IJETTCS) **3**(1), 75–79 (2014)

3. N. Adhikary, R. Shrivastava, A. Kumar, S.K. Verma, M. Bag, V. Singh, Battering keyloggers and screen recording software by fabricating passwords. Int. J. Comput. Netw. Inf. Secur. **4**(5), 13 (2012)
4. M. Alzomai, A. Jøsang, The mobile phone as a multi OTP device using trusted computing, in *2010 Fourth International Conference on Network and System Security* (IEEE, 2010, September), pp. 75–82
5. P. Ambalakat, Security of biometric authentication systems, in *21st Computer Science Seminar* (2005, April), p. 1
6. D. An, Find Out How You Stack Up to New Industry Benchmarks for Mobile Page Speed. Think with Google. 2018. https://www.thinkwithgoogle.com/marketing-resources/data-measurement/mobile-page-speed-new-industry-benchmarks/. Accessed 30 April 2018
7. E. Andreeva, Secret sharing in continuous access control system, using heart sounds, in *2012 XIII International Symposium on Problems of Redundancy in Information and Control Systems (RED)* (IEEE, 2012, September), pp. 5–6
8. A.J. Aviv, K.L. Gibson, E. Mossop, M. Blaze, J.M. Smith, Smudge attacks on smartphone touch screens. Woot **10**, 1–7 (2010)
9. W. Bao, H. Li, N. Li, W. Jiang, A liveness detection method for face recognition based on optical flow field, in *International Conference on Image Analysis and Signal Processing, 2009. IASP 2009* (IEEE, 2009, April), pp. 233–236
10. D.A. Buchanan, J. McCalman, *High Performance Work Systems: The Digital Experience* (Routledge, 2018)
11. A. Buriro, S. Gupta, B. Crispo, *Evaluation of Motion-Based Touch-Typing Biometrics for Online Banking* (2017)
12. Y. Canbay, M. Ulker, S. Sagiroglu, Detection of mobile applications leaking sensitive data, in *2017 5th International Symposium on Digital Forensic and Security (ISDFS)* (IEEE, 2017, April), pp. 1–5
13. I. Chingovska, A. Anjos, S. Marcel, On the effectiveness of local binary patterns in face anti-spoofing, in *Proceedings of the 11th International Conference of the Biometrics Special Interest Group* (No. EPFL-CONF-192369) (2012)
14. E.W.R. Chowdhury, M.S. Rahman, A.A. Al Islam, M.S. Rahman, Salty Secret: let us secretly salt the secret, in *2017 International Conference on Networking, Systems and Security (NSysS)* (IEEE, 2017, January), pp. 115–123
15. D. Coldewey, NIST declares the age of SMS-based 2-factor authentication over. TechCrunch. 2018. https://beta.techcrunch.com/2016/07/25/nist-declares-the-age-of-sms-based-2-factor-authentication-over/. Accessed 30 April 2018
16. A. Conklin, G. Dietrich, D. Walz, Password-based authentication: a system perspective, in *Proceedings of the 37th Annual Hawaii International Conference on System Sciences, 2004* (IEEE, 2004, January), 10 pp.
17. V. Conti, M. Collotta, G. Pau, S. Vitabile, Usability analysis of a novel biometric authentication approach for android-based mobile devices. J. Telecommun. Inf. Technol. (2014)
18. S.F. Darwaish, E. Moradian, T. Rahmani, M. Knauer, Biometric identification on android smartphones. Procedia Comput. Sci. **35**, 832–841 (2014)
19. M. De Marsico, C. Galdi, M. Nappi, D. Riccio, Firme: face and iris recognition for mobile engagement. Image Vis. Comput. **32**(12), 1161–1172 (2014)
20. Y. Durmus, K. Langendoen, Wifi authentication through social networks—a decentralized and context-aware approach, in *2014 IEEE International Conference on Pervasive Computing and Communications Workshops (PERCOM Workshops)* (IEEE, 2014, March), pp. 532–538
21. P. Elftmann, Secure alternatives to password-based authentication mechanisms. Laboratory for Dependable Distributed Systems, RWTH Aachen University (2006)
22. L. Fridman, S. Weber, R. Greenstadt, M. Kam, Active authentication on mobile devices via stylometry, application usage, web browsing, and GPS location. IEEE Syst. J. **11**(2), 513–521 (2017)
23. J. Galbally, S. Marcel, J. Fierrez, Image quality assessment for fake biometric detection: application to iris, fingerprint, and face recognition. IEEE Trans. Image Process. **23**(2), 710–724 (2014)

24. C.K. Goel, G. Arya, Hacking of passwords in windows environment. Int. J. Comput. Sci. Commun. Netw. **2**(3), 430–435 (2012)
25. J. Hu, L. Peng, L. Zheng, XFace: a face recognition system for Android mobile phones, in *2015 IEEE 3rd International Conference on Cyber-Physical Systems, Networks, and Applications (CPSNA)* (IEEE, 2015, August), pp. 13–18
26. S. Hussain, B.U.I. Khan, F. Anwar, R.F. Olanrewaju, Secure annihilation of out-of-band authorization for online transactions. Indian J. Sci. Technol. **11**(5), 1–9 (2018)
27. S.H. Islam, G.P. Biswas, A more efficient and secure ID-based remote mutual authentication with key agreement scheme for mobile devices on elliptic curve cryptosystem. J. Syst. Softw. **84**(11), 1892–1898 (2011)
28. A.K. Jain, A. Ross, S. Pankanti, Biometrics: a tool for information security. IEEE Trans. Inf. Forensics Secur. **1**(2), 125–143 (2006)
29. H.K. Jee, S.U. Jung, J.H. Yoo, Liveness detection for embedded face recognition system. Int. J. Biol. Med. Sci. **1**(4), 235–238 (2006)
30. J. Jeong, M.Y. Chung, H. Choo, Integrated OTP-based user authentication and access control scheme in home networks, in *Asia-Pacific Network Operations and Management Symposium* (Springer, Berlin, Heidelberg, 2007, October), pp. 123–133
31. B.U.I. Khan, A.M. Baba, R.F. Olanrewaju, S.A. Lone, N.F. Zulkurnain, SSM: secure-split-merge data distribution in cloud infrastructure, in *2015 IEEE Conference on Open Systems (ICOS)* (IEEE, 2015, August), pp. 40–45
32. B.U.I. Khan, R.F. Olanrewaju, A.M. Baba, A.A. Langoo, S. Assad, A compendious study of online payment systems: past developments, present impact, and future considerations. Int. J. Adv. Comput. Sci. Appl. **8**(5), 256–271 (2017)
33. B.U.I. Khan, R.F. Olanrewaju, F. Anwar, R.N. Mir, A.R. Najeeb, Scrutinizing internet banking security solutions. Special Issue on Multimedia Information Security Solutions on Social Networks (in press) (2018)
34. J.M. Kizza, *Ethical and Social Issues in the Information Age*, vol. 999 (Springer, 2007)
35. W.H. Lee, R. Lee, Implicit sensor-based authentication of smartphone users with smartwatch, in *Proceedings of the Hardware and Architectural Support for Security and Privacy 2016* (ACM, 2016, June), p. 9
36. G. Lovisotto, R. Malik, I. Sluganovic, M. Roeschlin, P. Trueman, I. Martinovic, Mobile biometrics in financial services: a five factor framework. Technical Report CS-RR-17–03, Oxford University (2017)
37. J. Määttä, A. Hadid, M. Pietikäinen, Face spoofing detection from single images using micro-texture analysis, in *2011 international joint conference on Biometrics (IJCB)* (IEEE, 2011, October, pp. 1–7
38. U. Mahbub, R. Chellappa, PATH: person authentication using trace histories, in *Ubiquitous Computing, Electronics & Mobile Communication Conference (UEMCON), IEEE Annual* (IEEE, 2016, October), pp. 1–8
39. S. Mare, A.M. Markham, C. Cornelius, R. Peterson, D. Kotz, Zebra: zero-effort bilateral recurring authentication, in *2014 IEEE Symposium on Security and Privacy (SP)* (IEEE, 2014, May), pp. 705–720
40. J. Marous, Millennials Are Leading the Digital Banking Revolution (2017). The Financial Brand. https://thefinancialbrand.com/64369/millennials-mobile-banking-digital-engagement-trends/. Accessed 30 April 2018
41. B.K. Marshall, Tips for Avoiding Bad Authentication Challenge Questions. White Paper (2007)
42. M. Masihuddin, B.U.I. Khan, M.M.U.I. Mattoo, R.F. Olanrewaju, A survey on e-payment systems: elements, adoption, architecture, challenges and security concepts. Indian J. Sci. Technol. **10**(20), 1–19 (2017)
43. S. McQuiggan, J. McQuiggan, J. Sabourin, L. Kosturko, *Mobile Learning: A Handbook for Developers, Educators, and Learners* (Wiley, 2015)
44. R. McWaters, *A Blueprint for Digital Identity* (World Economic Forum, 2016)
45. T. Mehraj, B. Rasool, B.U.I. Khan, A. Baba, A.G. Lone, Contemplation of effective security measures in access management from adoptability perspective. Int. J. Adv. Comput. Sci. Appl. **6**(8), 188–200 (2015)

46. W. Meng, W.H. Lee, S.R. Murali, S.P.T. Krishnan, Charging me and I know your secrets! towards juice filming attacks on smartphones, in *Proceedings of the 1st ACM Workshop on Cyber-Physical System Security* (ACM, 2015, April), pp. 89–98
47. W. Meng, D.S. Wong, S. Furnell, J. Zhou, Surveying the development of biometric user authentication on mobile phones. IEEE Commun. Surv. Tutor. **17**(3), 1268–1293 (2015)
48. M.S. Mir, M.B.A. Suhaimi, B.U.I. Khan, M.M.U.I. Mattoo, R.F. Olanrewaju, Critical security challenges in cloud computing environment: an appraisal. J. Theor. Appl. Inf. Technol. **95**(10), 2234–2248 (2017)
49. A. Narayanan, V. Shmatikov, Fast dictionary attacks on passwords using time-space tradeoff, in *Proceedings of the 12th ACM Conference on Computer and Communications Security* (ACM, 2005, November), pp. 364–372
50. N.C. Nguyen, O.J. Bosch, F.Y. Ong, J.S. Seah, A. Succu, T.V. Nguyen, K.E. Banson, A systemic approach to understand smartphone usage in Singapore. Syst. Res. Behav. Sci. **33**(3), 360–380 (2016)
51. W. Ockenden, AM—eBay suffers catastrophic data breach in hack attack 22/05/2014. abc.net.au. 2014. http://www.abc.net.au/am/content/2014/s4009539.htm. Accessed 30 April 2018
52. L. O'Gorman, Comparing passwords, tokens, and biometrics for user authentication. Proc. IEEE **91**(12), 2021–2040 (2003)
53. R.F. Olanrewaju, B.U.I. Khan, A. Baba, R.N. Mir, S.A. Lone, RFDA: reliable framework for data administration based on split-merge policy, in *SAI Computing Conference (SAI), 2016* (IEEE, 2016, July), pp. 545–552
54. R.F. Olanrewaju, B.U.I. Khan, M.M.U.I. Mattoo, F. Anwar, A.N.B. Nordin, R.N. Mir, Securing electronic transactions via payment gateways—a systematic review. Int. J. Internet Technol. Secur. Trans. **7**(3), 245–269 (2017)
55. R.F. Olanrewaju, B.U.I. Khan, M.M.U.I. Mattoo, F. Anwar, A.N.B. Nordin, R.N. Mir, Z. Noor, Adoption of cloud computing in higher learning institutions: a systematic review. Indian J. Sci. Technol. **10**(36), 1–19 (2017)
56. Online fraud happened hacking my icici bank credit card (2013). http://www.grahakseva.com/complaints/130310/online-fraud-happened-hacking-my-icici-bank-credit-card. Accessed 30 April 2018
57. A. Osseiran, J.F. Monserrat, P. Marsch (eds.), *5G Mobile and Wireless Communications Technology* (Cambridge University Press, 2016)
58. B.R. Pampori, T. Mehraj, B.U.I. Khan, A.M. Baba, Z.A. Najar, Securely eradicating cellular dependency for e-banking applications. Int. J. Adv. Comput. Sci. Appl. (IJACSA) **9**(2), 385–398 (2018)
59. G. Pan, L. Sun, Z. Wu, S. Lao, Eyeblink-Based Anti-Spoofing in Face Recognition from a Generic Web Camera (2007)
60. K.B. Raja, R. Raghavendra, M. Stokkenes, C. Busch, Multi-modal authentication system for smartphones using face, iris and periocular, in *2015 International Conference on Biometrics (ICB)* (IEEE, 2015, May), pp. 143–150
61. N.K. Ratha, J.H. Connell, R.M. Bolle, An analysis of minutiae matching strength, in *International Conference on Audio-and Video-Based Biometric Person Authentication* (Springer, Berlin, Heidelberg, 2001, June), pp. 223–228
62. C. Rathgeb, A. Uhl, A survey on biometric cryptosystems and cancelable biometrics. EURASIP J. Inf. Secur. **2011**(1), 3 (2011)
63. K. Rathi, S. Sawarkar, Finger print matching algorithm for android. Int. J. Eng. Res. Technol. (IJERT) **2**(10), 3819–3823 (2013)
64. A.S. Reid, Financial crime in the twenty-first century: the rise of the virtual collar criminal, in *White Collar Crime and Risk* (Palgrave Macmillan, London, 2018), pp. 231–251
65. A.A. Ross, K. Nandakumar, A.K. Jain, *Handbook of Biometrics* (Springer, US, 2008)
66. A.R. Sadeghi, T. Schneider, I. Wehrenberg, Efficient privacy-preserving face recognition, in *International Conference on Information Security and Cryptology* (Springer, Berlin, Heidelberg, 2009, December), pp. 229–244

67. B. Shebaro, O. Oluwatimi, E. Bertino, Context-based access control systems for mobile devices. IEEE Trans. Dependable Secure Comput. **12**(2), 150–163 (2015)
68. D.F. Smith, A. Wiliem, B.C. Lovell, Face recognition on consumer devices: reflections on replay attacks. IEEE Trans. Inf. Forensics Secur. **10**(4), 736–745 (2015)
69. Standards for Biometric Technologies. NIST. 2018. https://www.nist.gov/speech-testimony/standards-biometric-technologies. Accessed 30 April 2018
70. J. Téllez, S. Zeadally, *Mobile Payment Systems: Secure Network Architectures and Protocols* (Springer, 2017)
71. C.C. Teo, H.F. Neo, Behavioral fingerprint authentication: the next future, in *Proceedings of the 9th International Conference on Bioinformatics and Biomedical Technology* (ACM, 2017, May), pp. 1–5
72. C.J. Tsai, C.C. Peng, M.L. Chiang, T.Y. Chang, W.J. Tsai, H.S. Wu, Work in progress: a new approach of changeable password for keystroke dynamics authentication system on smart phones, in *2014 9th International Conference on Communications and Networking in China (CHINACOM)* (IEEE, 2014, August), pp. 353–356
73. Y. Yang, J.S. Sun, C. Zhang, P. Li, Retraining and dynamic privilege for implicit authentication systems, in *2015 IEEE 12th International Conference on Mobile Ad Hoc and Sensor Systems (MASS)* (IEEE, 2015, October), pp. 163–171
74. P. Zhang, Y. Pei, A technology of user access-control table and identity authentication based on USB in LAN, in *2010 International Conference on Biomedical Engineering and Computer Science (ICBECS)* (IEEE, 2010, April), pp. 1–3
75. Z. Zhao, Z. Dong, Y. Wang, Security analysis of a password-based authentication protocol proposed to IEEE 1363. Theoret. Comput. Sci. **352**(1–3), 280–287 (2006)
76. B.U.I. Khan, R.F. Olanrewaju, F. Anwar, Rehashing system security solutions in e-banking. Int. J. Eng. Technol. **7**(4), 4905–4910 (2018)
77. B.U.I. Khan, R.F. Olanrewaju, F. Anwar, M. Yaacob, Offline OTP based solution for secure internet banking access, in *IEEE Conference on e-Learning, e-Management and e-Services (IC3e)* (IEEE, 2018, November), pp. 167–172

Using Fuzzy Neural Networks Regularized to Support Software for Predicting Autism in Adolescents on Mobile Devices

Paulo Vitor de Campos Souza, Augusto Junio Guimaraes, Vanessa Souza Araujo, Thiago Silva Rezende and Vinicius Jonathan Silva Araujo

Abstract Intelligent decision tree-based software was built to aid in predicting the adolescent with autism traits. This application, which is obtained and operated on mobile devices, uses artificial intelligence and machine learning techniques to assign probabilities to people who pass the test in the application. The app has a knowledge base that assists in the prediction of autism in adolescents. This paper intends to demonstrate the feasibility of using fuzzy neural networks to assist in predicting the identification of autism traits, mainly supported by a system capable of generating fuzzy rules more cohesive than a decision tree. Therefore, this article proposes the insertion of an interpretive technique based on an extreme learning machine to deal with questions provided by users that seek to obtain more immediate answers, based on classification binary labels. The tests performed with the base achieved high levels of precision for the proposed model and support, making it a viable alternative for the efficient prediction of adolescents with autism.

Keywords Fuzzy neural network · Mobile devices · Autism prediction · Machine learning

P. V. de Campos Souza (✉)
CEFET-MG/Faculty Una of Betim, Belo Horizonte, Betim, Brazil
e-mail: goldenpaul@informatica.esp.ufmg.br

A. J. Guimaraes · V. S. Araujo · T. S. Rezende · V. J. S. Araujo
Faculty Una of Betim, Betim, Brazil
e-mail: augustojunioguimaraes@gmail.com

V. S. Araujo
e-mail: v.souzaaraujo@yahoo.com.br

T. S. Rezende
e-mail: silvarezendethiago@gmail.com

V. J. S. Araujo
e-mail: vinicius.j.s.a22@hotmail.com

M. Elhoseny and A. K. Singh (eds.), *Smart Network Inspired Paradigm and Approaches in IoT Applications*, https://doi.org/10.1007/978-981-13-8614-5_7

1 Introduction

Autism Spectrum Disorder (ASD) is a neurodevelopment condition that harms people to interact better with society, impairing their ability to learn, communicate, and speak. This type of illness, although present in people in the community, does not prevent it from having a quiet life, but better care about the way of learning, appropriate ways of dealing with the work environment are relevant to dealing with children and adolescent who have these learning disabilities. Several studies have been carried out for the application of intelligent techniques in the discovery and prediction of autism characteristics in adolescent, allowing smart systems to be created to assist in the diagnosis of people who have such cognitive definitions [1]. These techniques are intended to improve the diagnosis, treatment, and quality of health professionals' predictions. Regression techniques, rules-based decision trees, and standard classification techniques provided by WEKA software [2] were used. Decision tree techniques were employed in prediction techniques for the adolescent with or without autism, reaching between 94 and 99% of correct answers about how to correctly classify a patient with or without autism [3]. Recent research [4] have been carried out on different grouping forms of studies and inventive techniques to predict people with autism. In this context, a review of procedures to facilitate the collection of better methods was presented and consequently, a mobile application was developed [5] to promote and disseminate the diagnostic techniques for autism. The app that can be downloaded via mobile devices contains questions that help identify the characteristics of autism, becoming a popular form of data collection to form bases for training the algorithms and more usual for detection of this type of diagnosis. The application allows parents to see if the traces of this mental illness are in their adolescent [5]. These original structures facilitate decision-making in the rules structure, based on the set of training provided, creating a tree of paths for decision-making and therefore has excellent results in a classification of patterns. However, decision trees have some disadvantages, such as rectangular partition classification, may not work correctly with numerical attributes and may have a multi-leafed structure and many branches, making it possible for the set of If-then rules to make the algorithm if the parameters to be queried are at the end of the tree structure. Other factors that can stand out as elements that do not contribute to the stability of the decision trees are the linear, and perpendicular borders to the axes and the high sensitivity to small perturbations in the training set generate very different networks. To overcome these problems, this article proposes the use of a hybrid structure that has the interpretability of the rule-based fuzzy systems and the learning provided by artificial neural networks. The fuzzy neural networks are models that implement the union of the techniques of neural systems and neural network to realize the classification of binary patterns. In this algorithm, we highlight the regularization method based on resampling to select the characteristics most relevant to the model. Therefore, the proposed approach resembles the intelligent system schema proposed in [5], allowing faster training and network formation because it is based on extreme learning techniques. These techniques work with weights generation in the hidden layer in a random way, allowing the weights of the output layer to be generated by a method based on the pseudoinverse of Moore–Penrose [6]. Compared with back propagation-based methods, this algorithm becomes more efficient, and weights are updated in a single step. The paper is organized as follows: In section II are the present terms of the literature that make up the study carried out in this paper. Already in Sect. III, the fuzzy neural network model proposed to improve the interpretability of the mobile device for identification of autism is presented. In Sect. IV, the base data used by the mobile device, configuration, and test results are shown. Finally, in Sect. V, the conclusion about the experiments is presented to the reader of the paper.

2 Literature Review

2.1 Autistic Spectrum Disorder

Autistic Spectrum Disorder (ASD) is a disease associated with the human mind disorders that can impair their coexistence with people and ordinary daily activities. As their treatment costs are high and there is a delay in diagnosis by regular evaluation, it was proposed the use of intelligent techniques to facilitate the identification of adolescent with autism [1]. When the preliminary determination gives strong indications of autism in children or adolescent, the following impacts on the life of this youngster can be minimized with adequate treatment; however, it has always been difficult to obtain accurate data so that health professionals could diagnose the patient's efficient way [3, 4]. In recent times, research has been improved, and with the aid of computational resources, it is possible to collect relevant data for new inferences about the study of ADS in adolescents. In Fig. 1 are presented some symptoms that adults can perceive in children or adolescent and also some behavioral information, disorders, and characteristics associated with the disease.

2.2 Expert Systems for the Diagnosis of Autism

In the work of [2], a group of algorithms provided by the WEKA data mining tool [2] was used to evaluate the database of the first of four modules of the Generic Autism Diagnostic Observation Program (ADOS). On this basis, we used 16 data classification algorithms that use association rules, decision tree, and neighbor neighbors' algorithms to classify the model classes. The models obtained great accuracy of accuracy (between 99 and 100% in the test cases), allowing the approach used by neural networks is a viable way of predicting people with autism. In the work of [3], the tests were performed in groups of children and adolescents from 5 to 16 years. In this analysis, we also used the neural network algorithms provided by WEKA [2] and the models are chosen for the test were the same as those used in [2]. Results

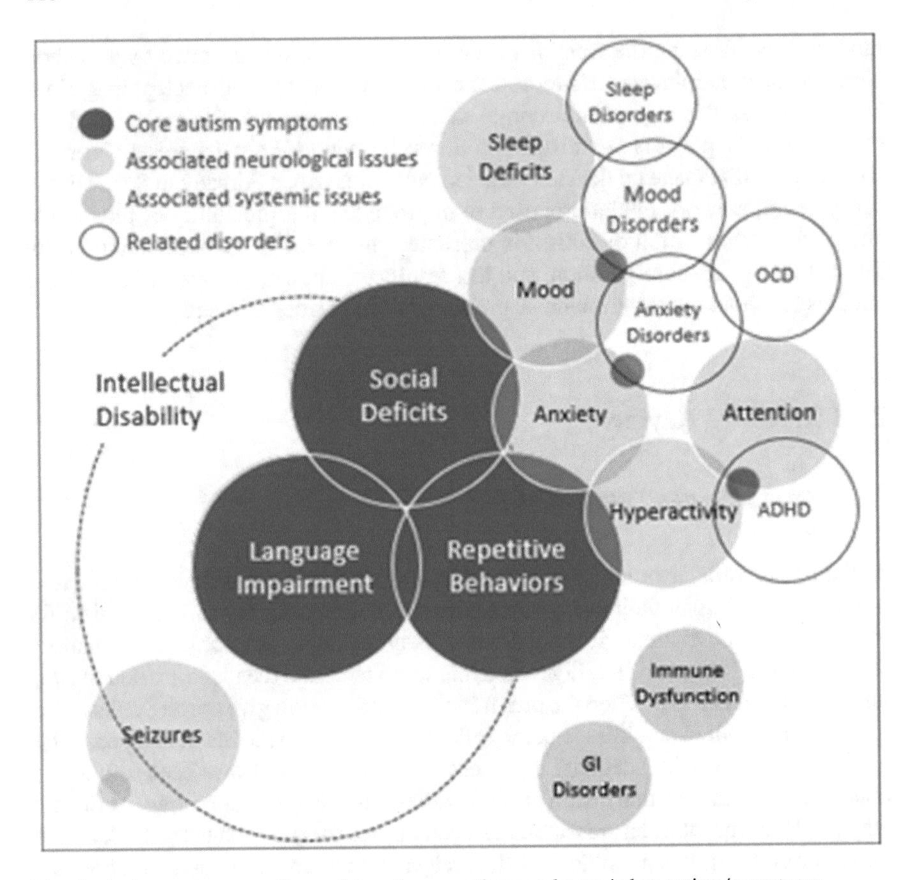

Fig. 1 Autism symptoms. *Source* https://www.autismspeaks.org/what-autism/symptoms

between 98 and 100% accuracy were also obtained in this analysis group, and it can be inferred that the approach of using neural networks also becomes feasible for analysis of young people with signs of autism. More recent studies [6] have evaluated autism in people of various characteristics. The results were still relevant, however, the best treatment on the basis of this work has drawn the attention of researchers such as [7] who, in addition to using the techniques of neural networks, criticized the methodological form in the works presented in [2, 3]. In [8] the use of four fuzzy neural networks assisted in prediction and comprehension in the prediction of autistic children. Finally, articles presented in [9, 10] seek to deal with the reduction of time and complexity in obtaining more precise answers to the patients' diagnosis. Figure 2 presents the performance of 15 machine learning algorithms evaluated for classifying autism cases and non-spectrum controls.

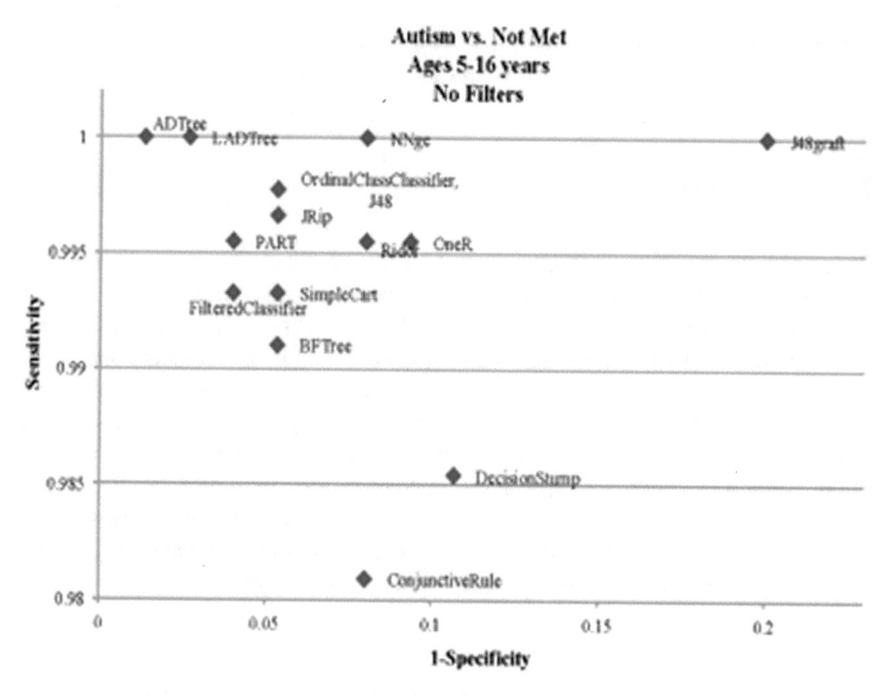

Fig. 2 Performance of 15 models in classifying autism [3]

2.3 *Techniques of Artificial Neural Networks and Fuzzy Systems*

In [7], an artificial neural network composed of an input layer, one or more hidden layers, and an output layer is defined. The network can be completely connected where each neuron is connected to all the neurons of the next layer, partially connected where each neuron is, or locally connected where there is a partial connection oriented to each type of functionality. To perform the training of a neural network, a set of data is required that contains patterns for training and desired outputs. In this way, the problem of neural network training is summarized in an optimization problem in which we want to find the best set of weights that minimizes the mean square error calculated between the network outputs and the desired outputs. Examples of artificial neural networks are the decision tree. They are commonly built by recursive partitioning. A univariate (single attribute) split is chosen for the root of the tree using some criterion (e.g., mutual information, gain-ratio, and Gini index). The data is then divided according to the test, and the process repeats recursively for each child. After a full tree is built, a pruning step is executed, which reduces the tree size [9]. Already the fuzzy systems are based on fuzzy logic, developed by [10]. His work was motivated because of the wide variety of vague and uncertain information in making human decisions. Some problems can be not solved with classical Boolean logic. In some situations, only two values are insufficient to solve a problem [11].

2.4 Fuzzy Neural Network

Fuzzy neural networks are characterized by neural networks formed of fuzzy neurons [12]. The motivation for the improvement of these networks lies in its easy interpretability, being possible to extract knowledge from its topology. These networks are formed by a synergistic collaboration between fuzzy set theory and neural networks allowing a wide range of learning abilities, thus providing models that integrate the uncertain information handling provided by the fuzzy systems and the learning ability conferred by the neural networks [11]. Thus, a fuzzy neural network can be defined as a fuzzy system that is trained by an algorithm provided by a neural network. Given this analogy, the union of the neural network with the fuzzy logic comes with the intention of softening the deficiency of each of these systems, making us have a more efficient, robust, and easy to understand a system [13].

2.5 A Mobile System for Autism Identification—Autism Spectrum Disorder Tests App

To facilitate the diagnosis of people with autism, a mobile device was developed in [5] to aid in application-directed questions whether or not a person is likely to have autistic characteristics. In this context, the software is downloaded free of charge from the mobile application stores and after choosing the language that the user will interact with the application, he begins to answer questions that were developed in studies of [1] to aid in the diagnosis of diseases. In Fig. 3, we can see how the software works and how it is fed by the people who respond to it all over the world. In Fig. 4, the interface presented to the user, the question styles, and the feedback received are highlighted.

2.6 Self-Organized Direction-Aware Data Partitioning Algorithm—SODA

The process by which fuzzy models treat data can determine how hybrid models can have the interpretability of their results closer to their real world. Models that are entirely data-driven are the targets of recent research and have achieved satisfactory results in cloud data cluster. This clustering concept focused on data is called Empirical Data Analytics (EDA) [14]. This concept brings together the data

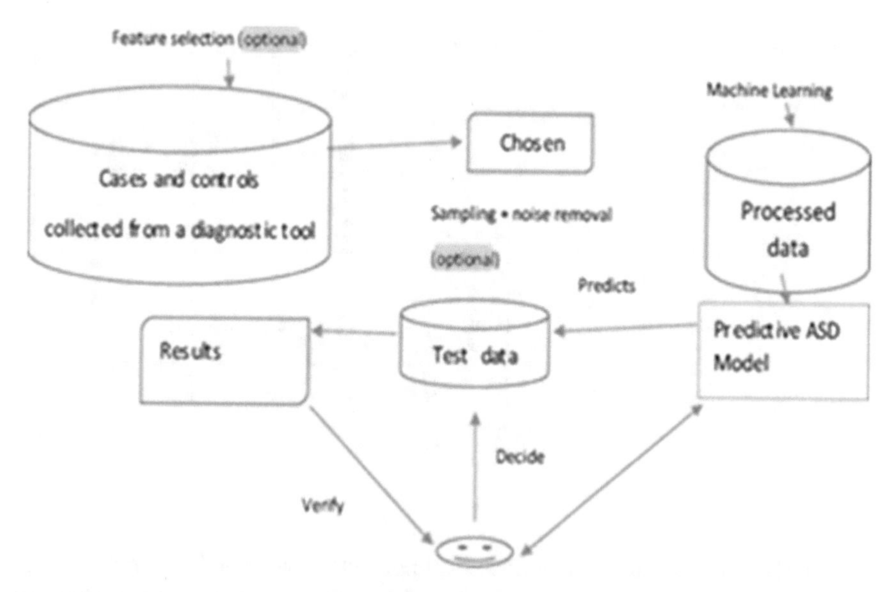

Fig. 3 Diagram of the decision system on autism [1]

Fig. 4 Mobile device interface for ASD identification [5]

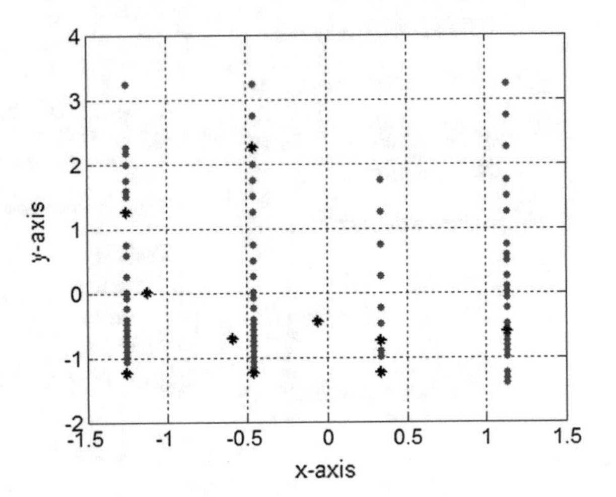

Fig. 5 SODA example

without statistical or traditional probability approaches, based entirely on the empirical observation of the input data of the model, without the need for any previous assumptions and parameters [15]. SODA is a data partitioning algorithm capable of identifying peaks/modes of data distribution and uses them as focal points to associate other aspects to data clouds that resemble Voronoi tessellation. Data clouds can be understood as a particular type of clusters, but with a much different variety. They are nonparametric, but their shape is not predefined and predetermined by the kind of distance metric used. Data clouds directly represent the properties of the local set of observed data samples [15]. The approach employs a magnitude component based on a traditional distance metric and a directional/angular segment based on the cosine similarity. The main EDA operators are described in [14], which are also suitable for streaming data processing. The EDA operators include the Cumulative Proximity, Local Density, and Global Density.

$$D_n(x_i) = \frac{\sum_{j=1}^{n} \pi_n(x_j)}{2n\pi_n(x_j)} \tag{1}$$

Figure 5 shows an example of the SODA definition and the center (black points) of density grouping defined by the algorithm.

2.7 Robust Activation Function—RAF

The activation functions allow small changes in weights and bias to cause only a slight change in output. Activation functions are a crucial element of artificial neural networks. They decide whether a neuron should be activated or not. That is, whether the information the neuron is receiving is relevant to the information provided or should be ignored. Overfitting caused by outliers and unreasonable selections of activation function and kernel function thus impairing the efficiency of algorithms performing tasks such as pattern classification [16]. The activation function is the nonlinear transformation along the input signal. This transformed output is then sent to the next layer of neurons as input. When we do not have the activation function, the weights and bias do a linear transformation. A linear equation is simple to solve but is limited in its ability to solve complex problems [16]. They are indispensable to give a representative capability for fuzzy neural networks by proposing a nonlinearity component. On the other hand, with this power, some difficulties appear, mainly due to the diversified variety of activation functions that can change the effectiveness of their activities according to specific properties of the database to which the model is being submitted. In general, by introducing nonlinear activation, the cost outside of the neuron is no large convex, making optimization more complicated. In problems that use parameterization by descent gradients, nonlinearity makes it more identifiable which elements need adjustment [17]. In models of fuzzy neural networks, the main functions of activation are those that use the hyperbolic tangent, sine, Gaussian, or linear. To avoid obstacles that can be created according to the variety of the data, the organization structure was developed based on the Gaussian activation function, a new proposal is as follows [18]: In the Gaussian activation function, if the uniformed sample $x_e = x/\|x\|_2$ and the output weight $w_e = w\|w\|_2$, then we can get the RAF activation function:

$$f_{RAF}(w; b; x) = \exp(-b(1 - \cos\theta), \tag{2}$$

where $\theta = < w; x >$ [18].

3 Fuzzy Neural Network for Autism Spectrum Disorder Test

Fuzzy Neural Networks (FNN) are neural networks formed by fuzzy neurons. These networks have as main characteristic the synergic collaboration between the theory of fuzzy sets and neural networks generating models that integrate the treatment of the uncertainty and interpretability provided by fuzzy systems and the learning ability provided by neural networks [11]. RNNs are composed of logical neurons, which are functional units that add relevant aspects of processing with learning capacity. They can be seen as multivariate nonlinear transformations between unit hypercubes [11]. Studies propose the generalization of logical neurons *and* and *or* that are constructed through extensions of *t-norms* and *s-norms*. One of the most important features of these new neurons, called *unineurons* [19] and *nullneurons* [20], are their ability to vary smoothly from a neuron to a neuron *or* to *and* and vice versa, depending on the need for the problem to be solved. This causes the final structure of the network to be determined by the training process, making this structure more general than neural networks formed only by classical logical neurons. These intelligent models have an architecture based on multilayered networks, where each of them has different functions in the model. In the works of [8, 13, 21–27], NFNs have three layers. In the [28] templates [29], its structure is composed of four layers. The function of each of these layers includes the concepts of fuzzy systems and artificial neural networks. In most models, the first layer is the one that partitions the input data, transforming them into nebulous logical neurons. Versions of fuzzy c-means, ANFIS, and clustering by the cloud are commonly applied. There are models responsible for pattern classification, fault prevention, and universal approximation. Already in training algorithms, we stand out models based on backpropagation, genetic, evolutionary models, and extreme learning machine. The way in which the fuzzification occurs in the fuzzy neural network determines its architecture, and a right approach, therefore, is capable of performing model with an optimized structure.

3.1 Network Architecture

The fuzzy neural network described in this chapter is composed of three layers. In the first layer, fuzzification is used through the concept of data density. The centers of the clusters are used to create the nebulous Gaussian neurons in the first layer. The wavelet transform defines the weights and bias of these neurons. Already in the second layer, the logical neurons of the andneuron type. These neurons have weights and activation functions determined at random and through t-norms and s-norms to aggregate the neurons of the first layer. To define the weights that connect the second layer with the output layer, the concept of a fast-learning machine is used to act on the neuron with a healthy activation function. Andneuron is used to construct fuzzy neural networks in the second layer to solve pattern recognition problems and bring interpretability to the model. Figure 6 illustrates the feedforward topology of the fuzzy neural networks considered in this paper. The first layer is composed of neurons whose activation functions are membership functions of fuzzy sets defined for the input variables. For each input variable x_{ij}, L clouds are defined A_{lj}, 1 = 1 L whose membership functions are the activation functions of the corresponding neurons. Thus, the outputs of the first layer are the membership degrees associated with the input values, i.e., $a_{jl} = \mu_l^A$ for $j = 1 \ldots N$ and $l = 1$ L, where N is the number of inputs and L is the number of fuzzy sets for each input results by SODA. The second layer is composed by L fuzzy andneurons. Each neuron performs a weighted aggregation of some of the first layer outputs. This aggregation is performed using the weights w_{il} (for $i = 1$ N and $l = 1$ L). For each input variable j, only one first layer output a_{jl} is defined as input of the lth neuron. So that **w** is sparse, each neuron of the second layer is associated with an input variable. Finally, the output layer is composed of one neuron whose activation functions are Raf. The output of the model is

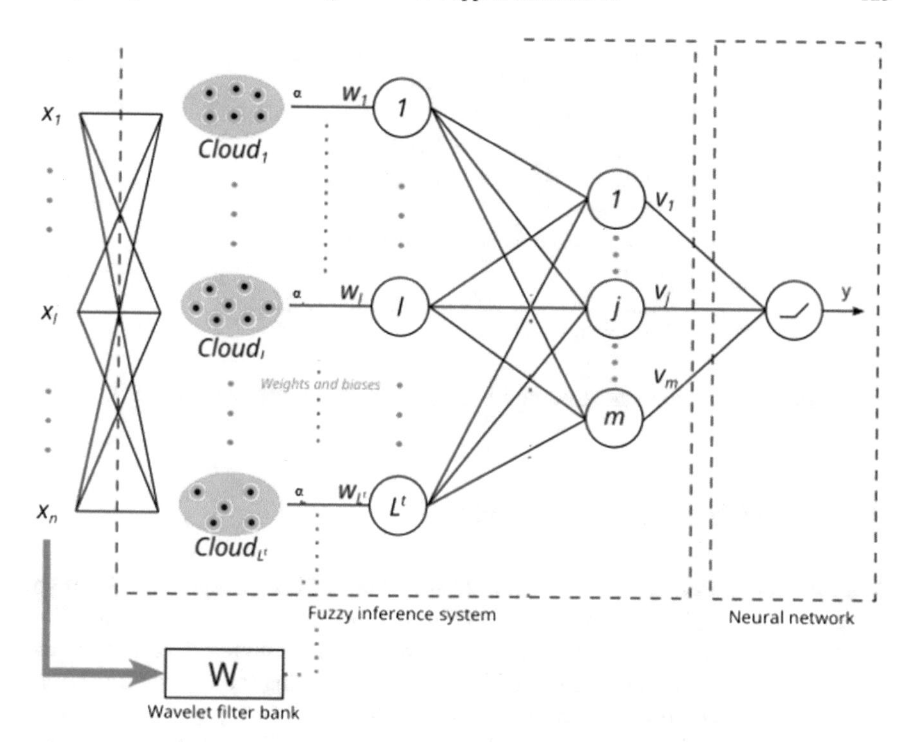

Fig. 6 FNN architecture

$$y = sign \sum_{j=0}^{l} f_{RAF}(z_l v_l) \tag{3}$$

where $z_0 = 1$, v_0 is the bias, and z_j and v_j, j = 1,..., l are the output of each fuzzy neuron of the second layer and their corresponding weight, respectively. Figure 4 presents an example of FNN architecture proposed in this paper.

Fuzzy logic neurons can be used to replace artificial neurons in original structures. The logical neurons used in the second layer of the model are of the andneuron type, where the input signals are individually combined with the weights and performed the subsequent global aggregation. The andneuron used in this work can be expressed as [11]

$$z = AND(w; z) = T_i^n = 1(w_i s x_i) \tag{4}$$

where T are *t-norms*, s is a *s-norms*. Fuzzy rules can be extracted from andneurons according to the following example:

Rule$_1$: If x_{i1} is A_1^1 with certainty w_{11}...
and $x_{i2}$is A_1^2 with certainty w_{21}...
Then y_1 is v_1
Rule$_2$: If x_{i1} is A_2^1 with certainty w_{12}...
and x_{i2} is A_2^2 with certainty w_{22}...
Then y_2 is v_2
Rule $_3$: If x_{il} is A_3^1 with certainty w_{13}...
Then y_3 *is* v_3
Rule$_4$: If x_{i2} is $A_3^2$with certainty w_{23}...
Theny_4 *is* v_4

These rules allow the creation of a building base for expert systems [13].

3.2 *Training Fuzzy Neural Network*

The membership functions in the first layer of the FNN are adopted as Gaussian, constructed through the centers obtained by the method of granularization of the SODA input space and by the randomly defined sigma. Another difference in the first layer is the definition of the fuzzy neuron weights using the wavelet transform. The number of neurons L in the first layer is defined according to the input data, and by the name of partitions (ρ), defined parametrically. This approach partitions the input space, following the definition logic of creating data nodes. The centers of these formed clouds make up the Gaussian activation functions of the fuzzy neurons. These changes will allow the adaptation of the data according to the basis submitted to the model, allowing a more independent and data-centered approach. The second layer performs the aggregation of the L neurons from the first layer through the andneurons.

After the construction of the L andneurons, the Bolasso algorithm [30] is executed to select LARS using the most significant neurons (called L_s). The final network architecture is defined through a feature extraction technique based on *l1* regularization and resampling. The learning algorithm assumes that the output hidden layer composed of the candidate neurons can be written as [22].

$$f(x_i) = \sum_{i=o}^{L_p} v_i z_i(x_i) = z(x_i)v \tag{5}$$

where $\mathbf{v} = [v_0, v_1, v_2, \ldots, v_L]$ is the weight vector of the output layer and $\mathbf{z}\ (x_i) = [z_0, z_1(x_i), z_2(x_i) z_L(x_i)\ ]$ the output vector of the second layer, for $z_0 = 1$. In this context, $\mathbf{z}\ (x_i)$ is considered as the nonlinear mapping of the input space for a space of nebulous characteristics of dimension $Ł_\rho$ [22].

LARS is a regression algorithm for high-dimensional data that is proficient in measuring exactly the regression coefficients but also a subset of candidate regressors to be incorporated in the final model. When judge a set of K distinct samples (x_i, y_i), where $x_i = [x_{i1}, x_{i2} \ldots x_{iN}]$ $\mathbb{R}$ and y_i $\mathbb{R}$ for $i = 1 \ldots K$, the cost function of this regression algorithm can be described as [22]

$$\sum_{i=1}^{K} x(z_i)v - y \| + \lambda \|_2 \|v\|_1 \tag{6}$$

where λ is a regularization parameter, commonly estimated by cross-validation. An efficient way of identifying which neurons are most activated for the problem is to verify through specific selection techniques using regression methods to which neurons are most relevant to a target problem. Insubstantial dimensional issues such as those of the pulsars, the selection of the best neurons allow the execution of the training to be more efficient, avoiding that unnecessary information is taken to the responses of the model. The first term of (6) corresponds to the sum of the squares of the residues (RSS). This term decreases as the training error decreases. The second term is an L_1 regularization term. Generally, this term is added, since it improves the generalization of the model, avoiding the super adjustment and can generate sparse models [31].

The LARS algorithm can be used to perform the model selection since for a given value of λ only a fraction (or none) of the regressors have corresponding nonzero weights. If $\lambda = 0$, the problem becomes unrestricted regression, and all weights are nonzero. As λ_{max} increases from 0 to a given value λ_{max}, the number of nonzero weights decreases to zero. For the problem considered in this paper, the z_l regressors are the outputs of the significant neurons. Thus, the LARS algorithm can be used to select an optimal subset of the significant neurons that minimize (6) for a given value of λ. Bolasso can be seen as a regime of consensus combinations where the most significant subset of variables on which all regressors agree when the aspect is the selection of variables is maintained [30]. Bolasso procedure is summarized in Algorithm 1.

Algorithm 1: Bolasso-bootstrap-enhanced least absolute shrinkage operator

(b1) Let n be the number of examples, (lines) in X.

(b2) Show n examples of (X, Y), uniformly and with substitution, called here ($Xsamp$, $ysamp$).

(b3) Determine which weights are nonzero given a λ value.

(b4) Repeat steps b1: b3 for a specified number of bootstraps b.

(b5) Take the intersection of the nonzero weights indexes of all bootstrap replications. Select the resulting variables.

(b6) Revise using the variables selected via non-regularized least squares regression (if requested).

(b7) Repeat the procedure for each value of b bootstraps and λ (actually done more efficiently by collecting interim results).

(b8) Determine "optimal" values for λ and b.

The LARS algorithm can be used to implement the model selection, considering, for a given value of, only a section (or none) of the regressors have same weights other than zero. If $\lambda = 0$, the problem becomes unrestricted regression, and all weights are nonzero. As λ_{max} increments from 0 to a given value λ_{max}, the amount of nonzero weights reduces to zero. For the problem admitted in this paper, the z_l regressors are the outputs of the important neurons [22].

Subsequently, following the determination of the network topology, the predictions of the evaluation of the vector of weights' output layer are performed. In this paper, this vector is considered by the Moore–Penrose pseudoinverse [22]:

$$v = Z^{+} y \tag{7}$$

Z^{+} is the Moore–Penrose pseudoinverse of z, which is the minimum norm of the least squares solution for the output weights. Where to remove the overfitting is one of the objectives of the ELM, it is possible to define the regularized model according to aim to find some appropriate w_i^{+}, b_i^{+} and h_i^{+} (i = 1,N^{+}) to satisfy [18]:

$$\begin{aligned} \|Z(w_i^{+}, \ldots, w_N^{+})v - y\| = min_{w_i, b_i, h_i} \\ \|Z(w_i, ..., w_N^{+}, b_i, \ldots b_N^{+})v - y\| \end{aligned} \tag{8}$$

Huang et al. [32] proved that as the weights and bias of the hidden layer are randomly defined the solution of the weights of the output layer can be found through the pseudoinverse.Meanwhile, expect to produce the weights with small values while avoiding overfitting. Thus, the ELM optimization modeling can be modeled mathematically as [18]

$$\begin{aligned} min_v \frac{1}{2}\|v\| + \frac{1}{2} C \sum_{i=1}^{N} \Xi_i^2 \\ s.t. \sum_{i=1}^{N^{+}} v f(w_i * x_j + b_j) - y_j = \Xi_j \end{aligned} \tag{9}$$

where C is a regularization factor normally used as 2^5 [32]. If the weights v are solved by Lagrange techniques after the transformation of Eq. (9) into an unrestricted optimization problem, we can solve the output layer weights of a model using ELM by the following equation [18]:

$$\begin{aligned} L_{delm} = \frac{1}{2}\|v\|^2 + C\frac{1}{2} \sum_{i=1}^{N} \Xi_i^2 - t_{ij} + \\ \sum_{i=1}^{N} \sum_{j=1}^{m} \alpha_{ij} (e(x_i) v_j \Xi_i j \end{aligned} \tag{10}$$

The KKT solution for this equation permits to find the equation that allows a better performance with training with a large number of samples.

$$f(x) = e(x)v = e(x) = \left(\frac{1}{c} + G^{+}G\right)^{-1} G^{+}y \tag{11}$$

Synthesized as demonstrated in Algorithm 2. It has three parameters:
1—the number of grid size, ρ;
2—the number of bootstrap replications, *bt*;
3—the consensus threshold, λ.

Algorithm 2: SODA-FNN training for autism

(1) Define grid size, ρ.
(2) Define bootstrap replications, bt.
(3) Define consensus threshold,
(4) Calculate L cluster in the first layer using SODA (ρ).
(5) Construct L fuzzy neurons with Gaussian membership functions constructed with center values derived from SODA and sigma defined at random.
(6) Define the weights and bias of the fuzzy neurons at random.
(7) Construct L andneurons with random weights and bias on the second layer of the network by welding the L fuzzy neurons of the first layer.
(8) **For all** K entries do
(8.1) Calculate the mapping $x_i(w_i)$
end for
(9) Select significant L_s using the lasso bootstrap according to the settings of *bt* and λ.
(10) Estimate the weights of the output layer (11)
(11) Calculate output **y** using Raf. (5)

4 Test of Classification Binary Patterns

4.1 Assumptions and Initial Test Configurations

The tests performed in this paper seek to find the accuracy of the fuzzy neural network model when classifying patterns of a pulsar base. The accuracy of the training and the test of the model will be realized through the conference of the values obtained by the model, comparing them with the expected result. When the result obtained is the same as the expected rating of the pulsar, a unit is counted. Therefore, the accuracy of the model is given in such a way as, to sum up, all the possible hits and to divide the value of this sum by the maximum possible number of hits for each stage of the model. This percentage is defined in 70% of the samples allocated for training and the remaining 30% for the test phase of the model. To avoid trends in the characteristics of each of the examples, a proposal was made where all the samples destined to training and test of the fuzzy neural network were randomly sampled. This ensures that there will be no dependencies of the data sequence for the model results. All samples involved in the test were normalized with mean zero and variance one. All outputs of the model were normalized to the interval [0,1] and the activation functions varied according to the test model. The values of bootstrap replications, the decision consensus are, respectively, 16 and 0.7. For fuzzy neural networks using equally spaced-apart perpendicular functions, we define 3, 5 Gaussian membership functions for each input variable. This number will be the same as the number of partitions that will address SODA in the data partition. These values were defined in previous cross-validated tests. To compare the performance of the fuzzy neural network model proposed in this article (SORAF-FNN) with the model proposed in [22] (ANFUNI-FNN) that uses unineurons and genfis1 to participate in the training data. The two models were compared with the same test criteria set out in [33] where the number of replicates is 500 times for each algorithm. To apply the same test conditions, the MLP, C4.5, and Naive Bayes (NB) algorithms were also used in the Weka software [2]. In the results, the table will be presented with the accuracy of the test, AUC, sensitivity, and the average time (in seconds) of execution of the 500 repetitions. The values in parentheses are the standard deviations. The test base reported in the previous item had its samples divided into training (70%) and test (30%). To avoid trends in the tests performed, all the available samples were exchanged and 30 measurements of the accuracy of each of the bases evaluated in each model analyzed were collected. The variables involved in the process were normalized with mean zero and unit variance. All outputs of the model were normalized to the range -1, 1. In all experiments, b = 16 and = 60% were considered as well as Gaussian in [22]. The variations occurred in the regularized version or not of the model and a variation of grid size (ρ) for the values 3, 5, 7, 9. The results of accuracy were presented in percentages and the standard deviation of the 30 repetitions of each experiment are in parentheses just ahead of the respective result of accuracy (training and test).

4.2 Database Used in the Tests

The original data collected have the characteristics presented in [4]. However, since the model works only with numerical attributes, a corresponding number has been assigned to each of the literal values. These attributes were replaced by numeric values without preferences or trends, sequentially starting from number 1, allowing for that attribute there is a context of equality between records that have the same number. Binary values were converted false to 0 and true to 1. Because it is only adolescent, the column that identified that identified was excluded from the trial because their values were unique and would not generate relevant to the model. The principal feature is age, gender, ethnicity, born with jaundice, family member with PDD, country, and ten questions.

Table 1 Accuracies of the model in the adolescent autism

Model	Acc.	AUC	Sens.	Time
SORAF	99.68 (0.99)	99.17 (0.77)	99.41 (1.02)	20.40 (7.12)
ANUNI	99.14 (1.68)	98.69 (0.02)	99.13 (0.42)	354.34 (16.87)
SVM	98.67 (1.65)	85.98 (0.76)	81.27 (2,24)	890.16 (93.32)
MLP	93.45 (4.56)	85.51 (3.18)	76.87 (3.45)	167.32 (28.56)
NB	93.33 (1.76)	78.98 (2.33)	64.54 (0.67)	267.12 (10.32)
C4.5	94.86 (4.06)	84.78 (11.76)	79.22 (1.54)	264.44 (36.43)

4.3 Autism Pattern Classification

In Table 1, it was found that the model proposed in this paper obtained the best accuracy results with a much shorter time than traditional models of literature and compared with the fuzzy neural network using the ANFIS in the input space partitioning.

4.4 Interpretability of the Problem Based on Fuzzy Rules

The data presented in Fig. 6 demonstrates the relationship in the Cartesian plane in results obtained from the mobile device (axis of ordinates) versus the age of the patients (abscissa axis). Also, consider that the elements in red are autistic people and the elements in blue are people without the presence of the disease. As in this example, five membership functions have also been generated for each of these input variables, we 10 can classify each as, for example, very low values, low, medium, high, and very high, which evaluated elements using words to distinguish them. In the cited example, we can elaborate fuzzy rules to aid in the construction of expert systems. If the age is very low (first membership function) and the note is too low (first membership function), the teen does not have autism. Similarly, if the age is too high (fifth membership function) and the note provided was high (fourth membership function), the teen has great chances of having autism (Fig. 7).

5 Conclusion

The tests performed verified that the use of a cloud data group and the Raf activation function is satisfactory for the classification of adolescent with autism executed by fuzzy neural networks. It has brought a more interpretive form to the problems of the autism, allowing the adjustment of parameters using the extreme learning makes possible the creation of systems based on fuzzy neural networks to act in the routine

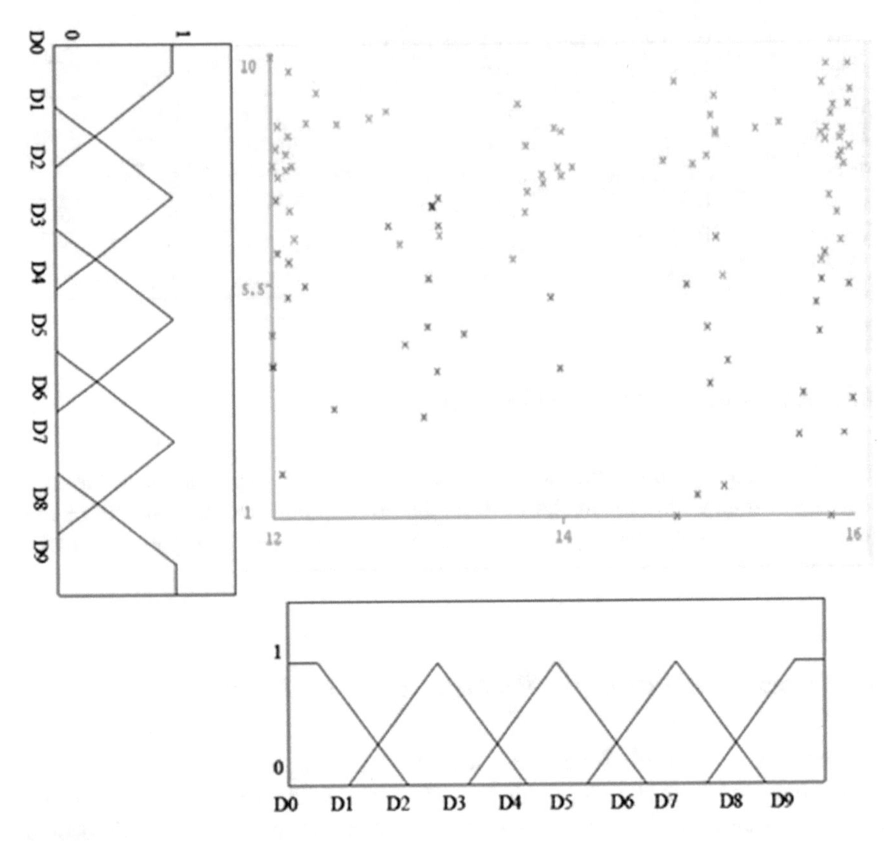

Fig. 7 Membership function in database

of researchers who work with problems related to the identification of problems in adolescent. The interpretability of the fuzzy neural network provides a better understanding of the factors that interfere in the identification of an adolescent with autism when compared to the approach of neural networks, which do not allow a more robust interpretation of the results. Although the cross-validation technique is a feasible approach to discover the optimal parameter of this type of problem, it is best to see it through optimization techniques. In this context, the use of fuzzy neural network becomes extremely effective and leaner than the decision five algorithms used in the literature. Based on these facts and presented tests, we can conclude that the fuzzy neural network model may help the mobile application developed in [5] to act with greater precision and less redundancy in decision-making about adolescents with autism. In future works, the models of fuzzy neural networks can be applied to verify the accuracy in the prediction of adolescents and adults with autism.

Acknowledgements The thanks of this work are destined to CEFET-MG and UNA.

References

1. F. Thabtah, in *Proceedings of the 1st International Conference on Medical and Health Informatics* (ACM, 2017), pp. 1–6
2. M. Hall, E. Frank, G. Holmes, B. Pfahringer, P. Reutemann, I.H. Witten, ACM SIGKDD Explor. Newsl. **11**(1), 10 (2009)
3. D. Wall, J. Kosmicki, T. Deluca, E. Harstad, V. Fusaro, Transl. Psychiatry **2**(4), e100 (2012)
4. F. Thabtah, Informatics for health and social care, 1–20 (2018)
5. F. Thabtah, Asdtests. A mobile app for asd screening (2017)
6. A. Albert, *Regression and the Moore-Penrose pseudoinverse* (Elsevier, Amsterdam, 1972)
7. A.D.P. Braga, A. Carvalho, T.B. Ludermir, *Redes neurais artificiais: teoria e aplicações* (Livros Técnicos e Científicos Rio de Janeiro, 2000)
8. A.P. Lemos, W. Caminhas, F. Gomide, in *Fuzzy Information Processing Society (NAFIPS), 2012 Annual Meeting of the North American* (IEEE, 2012), pp. 1–6
9. R. Kohavi, in *KDD*, vol. 96 (Citeseer, 1996), pp. 202–207
10. L.A. Zadeh, Inf. Control **8**, 3 (1965)
11. W. Pedrycz, IEEE Trans. Pattern Anal. Mach. Intell. (3), 289 (1991)
12. W. Pedrycz, F. Gomide, *Fuzzy Systems Engineering: Toward Human-centric Computing* (Wiley, New York, 2007)
13. W.M. Caminhas, H. Tavares, F.A. Gomide, W. Pedrycz, JACIII **3**(3), 151 (1999)
14. P.P. Angelov, X. Gu, J.C. Príncipe, IEEE Trans. Cybern. (2017)
15. X. Gu, P. Angelov, D. Kangin, J. Principe, Inf. Sci. **423**, 80 (2018)
16. M. Leshno, V.Y. Lin, A. Pinkus, S. Schocken (1992)
17. B. Karlik, A.V. Olgac, Int. J. Artif. Intell. Expert. Syst. **1**(4), 111 (2011)
18. S. Liu, L. Feng, Y. Xiao, H. Wang, Neurocomputing **144**, 318 (2014)
19. W. Pedrycz, IEEE Trans. Fuzzy Syst. **14**(6), 860 (2006)
20. M. Hell, P. Costa, F. Gomide, in *IEEE International Joint Conference on Neural Networks, 2008. IJCNN 2008. (IEEE World Congress on Computational Intelligence)* (IEEE, 2008), pp. 3653–3659
21. P.V.C. Souza, Int. J. Appl. Eng. Res. **13**(5), 2985 (2018)
22. P.V. de Campos Souza, G.R.L. Silva, L.C.B. Torres, in *2018 IEEE Conference on Evolving and Adaptive Intelligent Systems (EAIS)*, 2018, pp. 1–8. https://doi.org/10.1109/EAIS.2018.8397176
23. P.V. de Campos Souza, P.F.A. de Oliveira, in *2018 IEEE Symposium on Computer Applications and Industrial Electronics (ISCAIE)* (IEEE, 2018)
24. F. Bordignon, F. Gomide, Neurocomputing **127**, 13 (2014)
25. R. Ballini, F. Gomide, in *Proceedings of the 2002 IEEE International Conference on Fuzzy Systems, 2002. FUZZ-IEEE'02.*, vol. 1 (IEEE, 2002), pp. 785–790
26. P.V. de Campos Souza, L.C.B. Torres, in *Fuzzy Information Processing*, ed. by G.A. Barreto, R. Coelho (Springer International Publishing, Cham, 2018), pp. 13–23
27. R. Rosa, F. Gomide, R. Ballini, in *2013 12th International Conference on Machine Learning and Applications (ICMLA)*, vol. 2 (IEEE, 2013), pp. 378–383
28. Y. Shi, S. Jian, in *IOP Conference Series: Earth and Environmental Science*, vol. 128 (IOP Publishing, 2018), p. 012001
29. V.T. Yen, W.Y. Nan, P. Van Cuong, Neural Comput. Appl. 1–14 (2018)
30. F.R. Bach, in *Proceedings of the 25th International Conference on Machine Learning* (ACM, 2008), pp. 33–40
31. B. Efron, T. Hastie, I. Johnstone, R. Tibshirani et al., Ann. Stat. **32**(2), 407 (2004)
32. G.B. Huang, H. Zhou, X. Ding, R. Zhang, IEEE Trans. Syst., Man, Cybern. Part B (Cybernetics) **42**(2), 513 (2012)
33. D.P. Wall, R. Dally, R. Luyster, J.Y. Jung, T.F. DeLuca, PloS one **7**(8), e43855 (2012)

Performance Evaluation of Supervised Machine Learning Classifiers for Analyzing Agricultural Big Data

R. Anusuya and S. Krishnaveni

Abstract Big Data deals with huge volume of growing datasets, which are complex and having various autonomous sources. Storing and processing of such huge data was very tedious using earlier technologies so thus, the concept of Big Data came into existence. It was a tedious task for the end users to correctly identify the required data from a huge volume of unstructured data. The process of converting unstructured data into an organized and structured form, which will help the end user to easily retrieve the required data, is performed. So, applying classification techniques upon huge transactional database will provide required data to the end users from large datasets in a more simple way via Internet of Things (IoT). The two main categories of classification techniques are supervised and unsupervised techniques. In this paper, we have analyzed the performance of various supervised machine learning techniques on agricultural big dataset. Further, this paper shows the application of various analyzed techniques, its advantages, and limitations.

Keywords Big Data · Mining · Classification · IoT · Machine learning

1 Introduction

Developing IoT-enabled technologies and their solutions is a major challenge. However, IoT is about the pervasive collection and sharing of data toward a common goal [1]. Big Data has now become part of every sector and function of the global economy. Big Data is nothing but a collection of datasets that is too large and complex and cannot be accessed using conventional database software tools to search, manage, store, and process the data within a stipulated time period. The various challenges faced in Big Data analytics are data accessing and arithmetic computing procedures,

R. Anusuya · S. Krishnaveni (✉)
Department of Information Technology, Pioneer College of Arts and Science, Coimbatore, Tamil Nadu, India
e-mail: sss.veni@gmail.com

R. Anusuya
e-mail: anusuyayogesh@gmail.com

M. Elhoseny and A. K. Singh (eds.), *Smart Network Inspired Paradigm and Approaches in IoT Applications*, https://doi.org/10.1007/978-981-13-8614-5_8

semantics, and domain knowledge for different Big Data applications and the difficulties raised by Big Data volumes, distributed data distribution, and its complex dynamic characteristics. In order to face these challenges, the Big Data framework is divided into various levels [2].

Level I focuses on data accessing and arithmetic computing procedures. A suitable platform like Hadoop, R language is essential for computing the large volume of distributed dynamic information. Level II focuses on semantics and domain knowledge for different Big Data applications [3]. In this rapidly developing social network era, each and every user is linked to each other to share their knowledge. This knowledge is represented by group leader or communities as a model and so on, so in order to understand their semantics, application knowledge is essential for both low-level data access and for high-level mining algorithm designs [4–6]. Level III deals with the problems faced due to huge volume of distributed data and its dynamic characteristics [7]. The information is accumulated from different factors such as domestic and international news, government reports and natural disasters, and so on, hence, the user is not able to retrieve the required and appropriate over such a complex and voluminous data, and hence, it is crucial that such a data should be classified appropriately and presented to the user for his convenience and ease of access.

The abovementioned challenges can be solved using the classification technique, which can format and process the data according to their requirements. The processing is done by applying any one of the classification algorithms and analyzed to get the target system [8].

2 Classification Overview

Classification is a technique, which is used to classify unstructured data into the structured class and groups. This technique helps the user to discover the knowledge and process it according to their needs. It helps the user to take a correct decision [9]. Classification involves two phases like learning process and execution phase. The rules and patterns are generated in the learning phase and evaluation or testing of the given dataset is carried out in the second phase. The accuracy over a particular dataset can be achieved only when the second phase is properly evaluated and analyzed. This section briefly describes the supervised classification methods such as K-Nearest Neighbor, Naive Bayes, Decision Tree, and Support Vector Machine [10–12].

2.1 Supervised Methods

Supervised learning, also called as machine learning, is one of the classification methods, which group the data into instances. The processing involves two phases the training and testing phase [13]. In the training phase, the training set is a given a set of correctly classified instances, using this training set, the algorithm is designed to learn the instances. Then the same algorithm is used for prediction based on the trained data. This technique can be used when the user has the specific target value, which helps to predict the correct result from our data. The subset of training data along with its target value is essential to process and get the desired results.

Decision Tree (DT)

It is a hierarchical model, which can be used as a filter to handle the voluminous data. This method is one of efficient techniques, which gives best classification accuracy on all datasets. It automatically tunes itself between precision, which can train faster and can provide best result on unclassified data [14]. The features of the data are the instances, which are divided into discrete valued set of properties. This algorithm works faster and gives reasonably accurate results. The decision tree method involves the learning on large datasets by parallelizing process. In this method, decision trees are straightforwardly reduced into rules, which helps to break large dataset into n partitions, and then learn a decision tree on each of the n partitions in parallel. The decision tree becomes bigger when processing on n processor is done independently. In that case, they are combined to form an individual tree this process can be used in Meta-learning. Pruning is another technique, which can be used to remove the nodes which affects the accuracy of classification. This technique can be adopted when the training set is large. The pruning can be done using many methods. A pessimistic pruning method is adopted in C4.5, which can work fast and provides trees which can perform accurately. Thus, the main goal of Decision Tree is to find out the smallest tree that would make the data after split as pure as possible.

C4.5 Algorithm

C4.5 algorithm is an expansion of prior ID3 calculation, which can be used to manage both continuous and discrete properties, missing values, and prune trees after construction. The trees generated by this algorithm can be used for grouping, hence, they are called as statistical classifier. This algorithm generates decision trees from a set of training data, which are given in pairs: input object and the desired output value (class).

In C4.5 algorithm, the training set is analyzed and both training and test cases are accurately arranged by the classifier. The input object is the test example for which the output value is predicted. For example, let the training sample data S be represented as S1, S2, S3, …, Sn which is already trained. Each sample Si has a feature vector (x1,i, x2,i, …, xn,i) where xj denotes attributes or features of the sample and the class in which Si comes. C4.5 selects one attribute of the data from each node of the tree and splits samples into a set of subsets, which results in one class or the other. The condition is that to split is normalized information gain or the difference in entropy, which is a nonsymmetric measure of the difference between two probability distribution P and Q. The attribute which has the highest entropy value is chosen to make the decision. The steps adopted in C4.5 algorithm is listed below.

- Assume all the samples in the list belong to the same class. If it is true, it simply creates a leaf node for the decision tree so that particular class will be selected.
- None of the features provides any information gain. If it is true, C4.5 creates a decision node higher up the tree using the expected value of the class.
- An instance of previously unseen class is encountered. Then, C4.5 creates a decision node higher up the tree using the expected value

Support Vector Machine (SVM)

A machine learning technique used, nowadays, has the ability to make the computer learn and use algorithm or techniques to perform different tasks and activities efficiently. The main problem is that a technique is needed to represent the complex data and to remove the bogus data. Hence, SVM machine learning classifier is used for classification, which maps the original data into a higher dimension by classifying points into two disjoint half spaces. Here, a function is constructed to correctly predict the class to which the new points and old point belong.

In this Big Data era, the main problem is that maximum margin or separation is needed to make use of a decision boundary to classify the data accurately, this leads to a comparison of one dataset nearer to others. This situation is faced when the data is structured or linear but mostly, the data available for classification is unstructured/nonlinear and is inseparable. To get accurate classification results using this dataset, SVM kernels are used. The performance of the traditional classifiers degrades when the size of the data increases, however, SVM can be used to avoid this problem. SVM is the most promising technique, which can balance properly and accurately classify a huge amount of data. The complexity and error can be controlled explicitly.

The second advantage of using this SVM is that separate kernels can be used to deal with a particular problem that can directly deal with data without the need for a feature extraction process. In traditional methods when the feature extraction process is performed, a huge amount of data is lost.

Support Vector Machine (SVM) is one of the classification techniques, which can be used to process on large training data. The big and complex data can be processed using SVM because the result obtained from SVM is greatly influenced when the dataset is too noisy. The problem of overfitting, handling of huge volume of data, noisy data, and its complexity can be easily optimized using SVM. In this separate, kernels are used to reveal the largest eigenvalues, which are represented in the quantum form using corresponding eigenvectors of the training data [14].

ID3 Algorithm

ID3is a classifier in which calculation starts with the original set as the root hub. On every cycle of the algorithm, it concentrates on every unused attribute of the set and picks out the entropy or data pick up IG(A) of that attribute. Here, it chooses an attribute which has the smallest entropy or biggest data gain value. The algorithm proceeds to repeat recursively on each and every item in the subset by considering only items that are never selected before. Recursion on a subset comes to a halt in one of these cases:

- Every element in the subset belongs to the same class (+ or –), then the node is turned into a leaf and labeled with the class of the examples.
- If there are no more attributes to be selected but the examples still do not belong to the same class (same are + and some are –), then the node is turned into a leaf and labeled with the most common class of the examples in that subset.
- If there are no examples in the subset, then this happens when the parent set is found to be matching with a specific value of the selected attribute. For example, if there was no example matching with marks >= 100, then a leaf is created and is labeled with the most common class of the example in the parent set.

The algorithm works as follows,

- Calculate the entropy for each attribute using the dataset S.
- Split the set S into subsets using the attribute for which entropy is minimum (or, equivalently, information gain is maximum).
- Construct a decision tree node containing that attribute in a dataset.
- Recurse on each member of subsets using remaining attributes.

K-Nearest Neighbors Algorithm

The Nearest Neighbor (NN) rule differentiates the classification of unknown data point on the basis of its closest neighbor whose class is already known. M. Cover and P. E. Hart have proposed k-Nearest Neighbor (KNN) in which nearest neighbor is calculated on the basis of estimation of k, which indicates the number of nearest neighbors to be considered to characterize the class of a sample data point. In this method, more than one closest neighbor is utilized to determine the class in which the given data point belongs to. Hence, this algorithm is called as KNN. The data samples must be available in the memory at the run time and hence, it is also called as memory-based technique. T. Bailey and A. K. Jain have proposed an enhanced KNN algorithm, which focuses on weights. The training points are assigned with weights based on their distances from sample datapoint. The main drawback faced in this technique is the computational complexity and memory requirements to store a huge dataset. To overcome the memory limitation, the size of the data is decreased by using repeated patterns, which eliminates the additional data from the training dataset. To enhance over memory limit of KNNa, different systems of NN training dataset has to be organized. The implementation of this technique can be performed using k-d tree, ball tree, principal axis search tree, Nearest Feature Line (NFL), and orthogonal search tree. The tree structured training data is divided into nodes by using techniques like NFL, which uses a tunable metric to separate the training dataset according to planes. Using this algorithm, the speed of basic KNN algorithm can be increased. The algorithm works as follows:

K →number of nearest neighbors
For each object Xin the test set **do**
calculate the distance D(X,Y) between X and every object Y in the training set
neighborhood /the k neighbors in the training set closest to X
X.class → SelectClass (neighborhood)
End for

Naive Bayes Algorithm

The Naive Bayes Classifier technique works on the principle of Bayesian theorem and can be particularly used when the dimensionality of the inputs is high. The Bayesian Classifier is capable of calculating the most possible output based on the input. It also adds new raw data at runtime and have a better probabilistic classification rate. A naive Bayes classifier considers that the presence (or absence) of a particular feature (attribute) of a class is unrelated to the presence (or absence) of any other feature when the class variable is given. For example, a fruit may be considered to be an apple if it is red, round. Even if these features depend on each other or upon the existence of other features of a class, a naive Bayes classifier considers all of these properties to independently contribute to the probability that this fruit is an apple. The algorithm works as follows:

Bayes theorem provides a way of calculating the posterior probability, $P(c|x)$, from $P(c)$, $P(x)$, and $P(x|c)$.

$$P(c|\mathrm{X}) = P(x_1|c) \times P(x_2|c) \times \cdots \times P(x_n|c) \times P(c)$$

P(c|x) is the posterior probability of class (target) given predictor (attribute) of class.

- P(o) is called the prior probability of class.
- P(x|c) is the likelihood which is the probability of predictor of given class.
- P(x) is the prior probability of predictor of class.

ANN Algorithm

Artificial Neural Networks (ANNs) is one of the types of computer architecture, which deals with biological neural networks (Nervous systems of the brain) and are mainly used to approximate functions, which can depend on a large number of inputs and that are generally unknown. Artificial neural networks are represented as systems of interconnected "neurons" which can compute values from inputs and are capable of machine learning as well as pattern recognition due to their adaptive nature.

An artificial neural network works by creating connections between many different processing elements each corresponding to a single neuron in a biological brain. These neurons may be actually constructed or simulated by a digital computer system. Each neuron takes many input signals then based on an internal weighting produces a single output signal that is sent as input to another neuron. The neurons are strongly interconnected and organized into different layers. The input layer receives the input and the output layer produces the final output. In general, one or more hidden layers are sandwiched in between the two. This structure makes it impossible to forecast or know the exact flow of data.

Artificial neural networks typically start to work out with randomized weights for all their neurons. This means that initially, they must be trained to solve the particular problem for which they are proposed. A backpropagation ANN is trained by humans to perform specific tasks. During the training period, the correct ANN's output is computed by observing the pattern. Based on the correctness, the neural weightings that produced that output are reinforced; if the output is incorrect, those weightings responsible can be reduced.

3 Comparative Study

In this section, comparative study of Decision Tree, ID3, KNN, Naive Bayes, SVM, and Artificial Neural network has been done based on predictive accuracy, fitting speed, prediction speed, memory usage, area under curve, etc. The comparison is listed below in Table 1. Further, Table 2 shows its advantages, limitations, and applications of all these techniques.

Table 1 Comparison between supervised learning models

Performance measure	Naive Bayes	KNN	ID3	ANN	C4.5	SVM
Predictive accuracy	Low	Low	Low	High	Low	High
Training speed	Fast	Fast	Fast	Slow	Fast	Medium
Prediction speed	Depends on number of instance	Fast	Fast	Fast	Fast	Fast
Memory usage	High	High	High	High	Low	Low
Easy to interpret	No	No	Yes	Yes	Yes	Yes
Area under the curve	More	More	More	More	more	Less
Handles categorical predictors	Yes	Yes	Yes	No	Yes	No

Table 2 Advantages and disadvantages of classification algorithm

S. no.	Classification algorithm	Advantage	Disadvantage
1	K-Nearest Neighbor	• Zero cost of learning process • Classes are not linearly separable • Robust with regard to noisy training data • Best suited for multimodal learning	• Time consuming—for large dataset • More sensitive to noise and irrelevant data
2	Naive Bayes	• Simple to implement • Greatest computational efficiency and classification rate	• Performance depends on the number of dimensions used • Precision of the algorithm decreases if the amount of data is less
3	ID3 algorithm	• Produces more accurate results than KNN • Detection rate is increased • Consumption is reduced	• Requires large searching time • Generates long rules which is difficult to prune • Requires more memory
4	C4.5 algorithm	• Easy to implement • It can use both discrete and continuous values • Deals with noise • Easy to implement by building models	• Over fitting • Does not work well on small training dataset
5	Support vector machine (SVM)	• High accuracy • Works on non linearly separable data in the feature space	• Speed and size requirement in both training and testing is more • High complexity
6	Artificial neural network (ANN)	• Applicable to wide range of problems in real life • Can be used with a very few parameters • Easy to implement	• Requires high processing time • Learning can be slow

On comparison it can be concluded that SVM has higher prediction speed and good memory usage. On using correct kernel function, the difficulty to classify the data can be reduced. Using linear kernel, all applications can be interpreted easily.

4 Literature Review

Survey of different data mining algorithms used in the field of agriculture have been analyzed and some of them have been listed below:

S. no.	Title of the paper	Results and discussion	Algorithm applied	Authors
1	An investigation into the application of neural networks, fuzzy logic, genetic algorithms, and rough sets to automated knowledge acquisition for classification problems	Classifying soil in combination with GPS	K-means	I. Jagielska, C. Mattehews, T. Whitfort [15]
2	High resolution continuous soil classification with morphological soil profile descriptions	Prediction of yield in agriculture	Fuzzy set	K. Verheyen, D. Adriaens, M. Hermy, S. Deckers [16]
3	Crop productivity mapping based on decision tree and Bayesian classification	Scale back procedure on crop yielding using k-nearest neighbor algorithms	KNN	S. Veenadhari [17]
4	Data Mining Techniques in Agricultural and Environmental Sciences	Classifying the weather sample information into linearly severable	Support Vector Machine	A. Chinchulunn, P. Xanthopoulos, V. Tomaino, P. M. Pardalos [18]
5	Data mining Techniques for Predicting Crop Productivity	kharif and rabi crops production effected by climatic factors	Decision Trees	S. Veenadhari, Dr. B. Misra, Dr. C. D. Singh [19]
6	Unsupervised neural network approach to medical data mining techniques	Daily precipitations simulation of weather and other conditions	KNN	D. Shalvi, N. De Claris [20]
7	A K-nearest neighbor simulator for daily precipitation and other weather variable	Conducting climate impact studies	Support Vector Machine	B. Rajagopalan, U. Lal [21]
8	A vision-based hybrid classifier for weeds detection in precision agriculture through the Bayesian and Fuzzy k-Means paradigms	Wine fermentation	K-means	A. Tellaeche, X. P. BurgosArtizzu, G. Pajares, A. Ribeiro [22]

(continued)

(continued)

S. no.	Title of the paper	Results and discussion	Algorithm applied	Authors
9	Modeling and prediction of rainfall using artificial neural network and ARIMA techniques	Prediction of rainfall	Neural Networks	V. K. Somvanshi, et al. [23]
10	Downscaling of precipitation for climate change scenarios: a support vector machine approach	Using Bayesian network learning method developing the model for agriculture	Bayesian network	S. Tripathi, V. Srinivas, R. S. Nanjundiah [24]
11	Using data mining techniques to predict industrial wine problem fermentations	weeds precision detection in agriculture	Fuzzy set	A. Urtubia, J. R. Pérez Correa, A. Soto, P. Pszczolkowski [25]
12	Classification and Prediction of Future Weather by using Back Propagation Algorithm An Approach	Focuses on weather forecasts	Neural Networks	S. D. Sawaitul, Prof. K. P. Wagh, Dr. P. N. Chatur [3]
13	Analysis of Big Data technologies for use in agro-environmental science	Research in the agro-environmental domain has to deal with large and very diverse datasets, both in content, structure, and storage format	Querying method in agro-environmental domain	Xindong Wu, Xingquan Zhu, Gong-Qing Wu, and Wei Ding [26]

5 Dataset Construction

The objective of proposed work is to analyze the agriculture data using all mentioned supervised data classification techniques. This work aims to identify the classifier, which gives better performance using the same dataset. In this proposed work, agriculture data has been collected from the following sources.

Dataset in the agricultural sector, crop-wise agriculture data, agriculture data of different districts, agriculture data based on weather, temperature, and relative humidity.

Input dataset consist of 10-year data with the following parameters, namely year, area—different states within India.

The District, Year, Season, Crop, Area, Production, Productivity, Crop Label, Rainfall, Temperature, crop (paddy, cotton, groundnut, jowar, rice, and wheat.), season (kharif, rabi, and summer), area (in hectares), production (in tonnes), average

	A	B	C	D	E	F	G	H	I	J	K	L
1		District	Year	Season	Crop	Area	Productior	roductivit	CropLabel	Rainfall	emperature	
2	0	AMRAVAT	2001	Kharif	Bajra	1000	400	0.4	1	0	3	
3	5	AMRAVAT	2002	Kharif	Bajra	800	500	0.625	2	1	3	
4	10	AMRAVAT	2003	Kharif	Bajra	800	400	0.5	1	1	3	
5	15	AMRAVAT	2004	Kharif	Bajra	700	400	0.571429	2	1	2	
6	19	AMRAVAT	2005	Kharif	Bajra	300	100	0.333333	1	2	4	
7	22	AMRAVAT	2006	Kharif	Bajra	200	100	0.5	1	2	2	
8	25	AMRAVAT	2007	Kharif	Bajra	2	1	0.5	1	2	3	
9	29	AMRAVAT	2008	Kharif	Bajra	300	200	0.666667	2	1	3	
10	34	AMRAVAT	2009	Kharif	Bajra	200	200	1	3	1	3	
11	38	AMRAVAT	2010	Kharif	Bajra	300	200	0.666667	2	2	3	
12	43	AMRAVAT	2011	Kharif	Bajra	300	200	0.666667	2	2	4	
13	48	AMRAVAT	2012	Kharif	Bajra	300	200	0.666667	2	2	4	
14	59	AURANGA	2001	Kharif	Bajra	137400	86700	0.631004	2	0	2	
15	64	AURANGA	2002	Kharif	Bajra	149900	131100	0.874583	2	0	2	
16	69	AURANGA	2003	Kharif	Bajra	128700	110200	0.856255	2	1	2	
17	74	AURANGA	2004	Kharif	Bajra	128900	107900	0.837083	2	1	1	
18	79	AURANGA	2005	Kharif	Bajra	131400	150000	1.141553	3	1	2	
19	82	AURANGA	2006	Kharif	Bajra	123100	139400	1.132413	3	2	1	
20	86	AURANGA	2007	Kharif	Bajra	1111	1362	1.225923	3	1	2	
21	91	AURANGA	2008	Kharif	Bajra	109100	103500	0.948671	3	1	2	
22	97	AURANGA	2009	Kharif	Bajra	122400	121800	0.995098	3	1	1	
23	102	AURANGA	2010	Kharif	Bajra	89800	126500	1.408686	4	1	1	
24	107	AURANGA	2011	Kharif	Bajra	68500	87800	1.281752	4	1	1	
25	112	AURANGA	2012	Kharif	Bajra	68500	44500	0.649635	2	0	2	

Fig. 1 Crop yield classification dataset

temperature (°C), average rainfall (mm), soil type, minimum rainfall required, and minimum temperature required.

Agri dataset 1 data consists of the following attributes like District, Year, Month, Rainfall, Average Rainfall, Temperature, Average Temperature, Pressure, and Average Pressure–Drought condition—using these attributes, the area which is affected by drought and not affected by drought based on temperature and rainfall has been classified. The result shows the classification of drought and drought-less areas.

Agri dataset 2 data consists of the following attributes like district, year, season, crop, area, productivity, crop label, average rainfall, and average temperature—using these attributes based on the crop label rainfall, season, and temperature, it classifies the crop to be grown in that season. The sample of both the dataset is given in Figs. 1 and 2. This dataset is organized in Excel Sheet and then converted as CSV extension for processing. The size of the both the datasets was 2.64 and 3.25 MB.

6 Performance Evaluation

The training set and the test data which needs to be classified have to categorized into two classes, so that SVM classifier can build a model that which can suit one of the two categories. Extraction of data from huge training set is modeled as a multidimensional classification problem with one class for each action and its aim is to assign a class label to a given action or activity. Phyton is used to evaluate the

	District	Year	Month	Rainfall	Average Rainfall	Temperature	Average Temperature	Pressure	Average Pressure	Drought Classification
0	Nagpur	2001	January	1	1	6	8	1	1	0
1	Nagpur	2001	February	1	1	6	8	2	2	0
2	Nagpur	2001	March	1	1	5	7	3	3	0
3	Nagpur	2001	April	1	1	3	4	4	4	0
4	Nagpur	2001	May	1	1	2	3	4	5	0
5	Nagpur	2001	June	5	3	1	2	4	4	0
6	Nagpur	2001	July	4	6	2	2	3	3	0
7	Nagpur	2001	August	5	5	2	2	3	3	0
8	Nagpur	2001	Septembe	2	3	2	4	3	3	0
9	Nagpur	2001	October	2	1	4	6	3	3	0
10	Nagpur	2001	Novembe	1	1	6	8	2	2	0
11	Nagpur	2001	Decembei	1	1	6	8	2	1	0
12	Nagpur	2002	January	1	1	6	8	2	1	1
13	Nagpur	2002	February	1	1	5	8	2	2	0
14	Nagpur	2002	March	1	1	5	7	3	3	0
15	Nagpur	2002	April	1	1	3	4	4	4	1
16	Nagpur	2002	May	1	1	2	3	5	5	1
17	Nagpur	2002	June	5	3	1	2	4	4	1
18	Nagpur	2002	July	2	6	1	2	3	3	0
19	Nagpur	2002	August	6	5	2	2	3	3	1
20	Nagpur	2002	Septembe	3	3	2	4	3	3	1
21	Nagpur	2002	October	1	1	5	6	2	3	1
22	Nagpur	2002	Novembe	1	1	6	8	2	2	1
23	Nagpur	2002	Decembei	1	1	6	8	1	1	1

Fig. 2 Drought classification dataset

performance of all the abovementioned classifiers on our both agricultural dataset and more accurate classification results have been obtained using tenfold cross validation. The performance metrics like Accuracy, Mean deviation, and Error rate has been evaluated to analyze all the classifiers. The confusion matrix is also called the contingency table (Fig. 3).

The accuracy and error rate is calculated as

- Nt—number of testing examples (a + b + c + d)
- N c—number of correctly classified testing examples (a + d)
- Classification accuracy: Acc $= \frac{Nc}{Nt}$
- (Misclassification) Error: It is the percentage of incorrect classification. E $= \frac{Nt-Nc}{Nt}$

The performance of the investigated methods is evaluated using an overall accuracy (ACC) and confusion matrix, where ACC corresponds to the proportion of overall correctly classified evaluation data points. The confusion matrix consists of True Positive Rate (TPR), True Negative Rate (TNR), False Positive Rate (FPR), and False Negative Rate (FPR).

TPR is defined as the coefficient of true positive classified points to the sum of true positive and false negative points, while TNR is the coefficient of true negative points to the sum of true negative and false positive points (Tables 3 and 4, Figs. 4, 5, and 6).

In both cases of SVM, the random choice of parameters yields a classification more accurate than the default classification in about 99% of the cases. Based on

Fig. 3 Confusion matrix to evaluate performance metrics

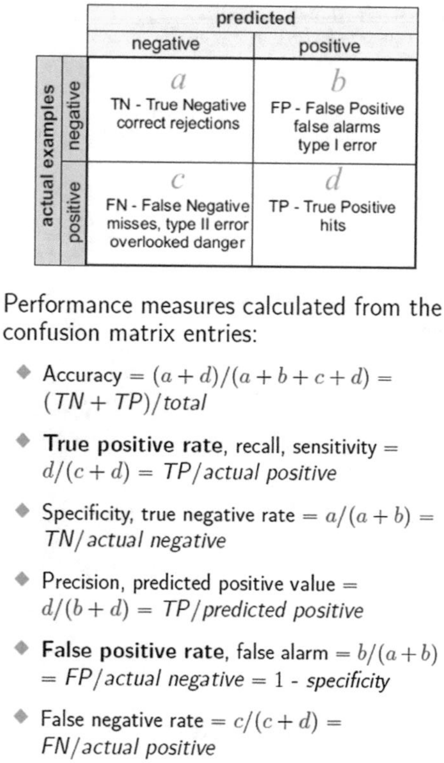

		predicted negative	predicted positive
actual examples	negative	a TN - True Negative correct rejections	b FP - False Positive false alarms type I error
actual examples	positive	c FN - False Negative misses, type II error overlooked danger	d TP - True Positive hits

Performance measures calculated from the confusion matrix entries:

- Accuracy $= (a + d)/(a + b + c + d) = (TN + TP)/total$
- **True positive rate**, recall, sensitivity $= d/(c + d) = TP/actual\ positive$
- Specificity, true negative rate $= a/(a + b) = TN/actual\ negative$
- Precision, predicted positive value $= d/(b + d) = TP/predicted\ positive$
- **False positive rate**, false alarm $= b/(a + b) = FP/actual\ negative = 1$ - *specificity*
- False negative rate $= c/(c + d) = FN/actual\ positive$

Table 3 Classifier performance using Agri dataset 1

Classifier	Mean accuracy (Acc)	Average error rate (E)	Average TPR	Average TNR	Time taken (s)
C4.5	83.596	0.54	0.61	0.72	1.4977
ID3	72.45	0.65	0.44	0.63	2.507
KNN	93.47	0.78	0.53	0.68	2.076
Naive Bayes	61.96	0.66	0.57	0.69	1.4977
SVM	96.95	0.36	0.64	0.87	0.5237

Table 4 Classifier performance using Agri dataset 2

Classifier	Mean accuracy (Acc)	Average error rate (E)	TPR	TNR	Time taken(s)
C4.5	72.56	0.58	0.53	0.68	1.567
ID3	67.32	0.64	0.46	0.71	2.034
KNN	89.23	0.87	0.43	0.73	2.076
Naive Bayes	61.54	0.54	0.55	0.56	1.230
SVM	90.89	0.46	0.67	0.78	0.4523

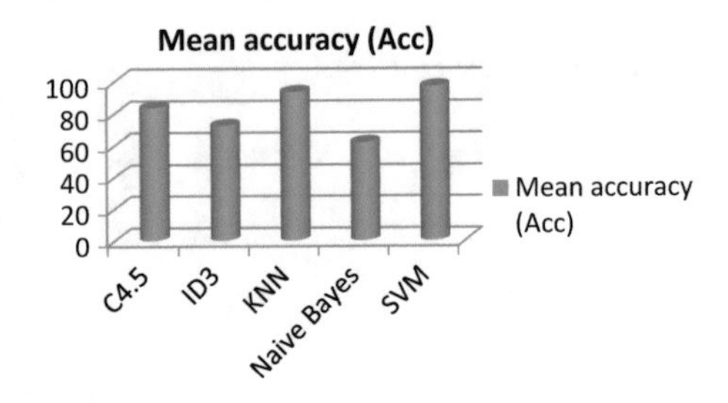

Fig. 4 Chart comparison of algorithm based on accuracy

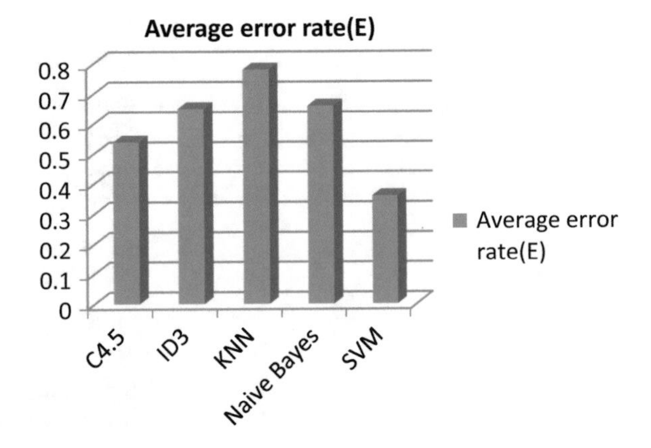

Fig. 5 Comparison based on mean error rate

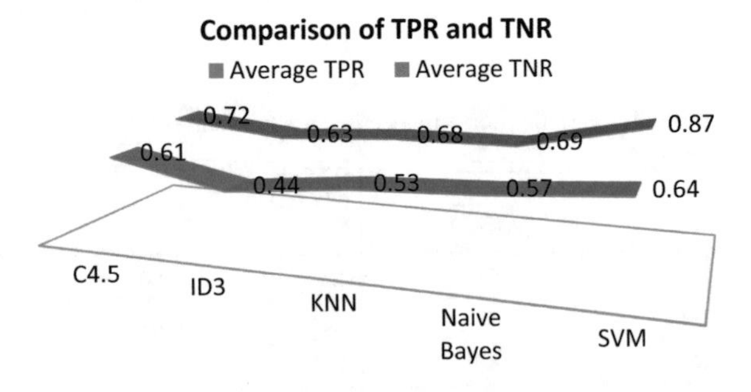

Fig. 6 Comparison based on average TPR and TNR

the training dataset, it is concluded that the average accuracy of SVM classifier is 0.9695. In the case C4.5, KNN, ID3, and Naive Bayes accuracy rate is 0.8954, 0.93857, and 0.6196, it indicates the low level. So, automatically, SVM classifier classified the dataset in a higher sense. SVM have high training performance and low generalization error, which pointed out the potential problems of SVMs when the training set is noisy and imbalanced.

Drought conditions differed among cities with regard to temperature, average rainfall, and average pressure. The crop that can be grown in the specific area for a particular temperature and rainfall has also been classified. The spectral analysis was sufficiently sensitive to capture the variation in temperature and rainfall between the different cities during the stipulated month.

7 Conclusion

In this paper, different supervised classification techniques on the era of agricultural Big Data have been implemented for analyzing its performance. All techniques being considered is better suited than the other for different applications. On comparing the performance of agricultural dataset, SVM performs well in terms of accuracy, error rate, time taken, and average TPR and TNR values. This paper also summarizes the advantages and disadvantages of the different classification techniques. The comparative results provided in this paper can be used to organize all kinds of user needs. Each technique has its uniqueness in terms of different metrics like accuracy, processing time, error rate, etc. This evaluation study indicates that the classification accuracy of SVM algorithm is best in terms of all the metrics on two datasets also, hence in future, we have an idea to enhance the performance of SVM algorithm so that it can still perform better for big data with less computational time.

References

1. S.K. Lakshmanaprabu, K. Shankar, A. Khanna, D. Gupta, J.J. Rodrigues, P.R. Pinheiro, V.H.C. De Albuquerque, Effective features to classify big data using social Internet of Things. IEEE Access **6**, 24196–24204 (2018)
2. N.G. Yethiraj, Applying data mining techniques in the field of agriculture and allied sciences. Int. J. Bus. Intell. **01**(02) (2012, December). ISSN: 2278-2400
3. S.D. Sawaitul, K.P. Wagh, P.N. Chatur, Classification and prediction of future weather by using back propagation algorithm—an approach. Int. J. Emerg. Technol. Adv. Eng. **2**(1), 110–113 (2012, January)
4. S.K. Lakshmanaprabu, K. Shankar, D. Gupta, A. Khanna, J.J.P.C. Rodrigues, P.R. Pinheiro, V.H.C. de Albuquerque, Ranking analysis for online customer reviews of products using opinion mining with clustering. Complexity, **2018**, Article ID 3569351, 9 (2018). https://doi.org/10.1155/2018/3569351
5. K. Shankar, S.K. Lakshmanaprabu, D. Gupta, A. Maseleno, V.H.C. de Albuquerque, Optimal feature-based multi-kernel SVM approach for thyroid disease classification. J. Supercomput. (2018). https://doi.org/10.1007/s11227-018-2469-4
6. M. Muslihudin, R. Wanti, N. Hardono, K. Shankar, M. Ilayaraja, A. Maseleno, D.R.M. Fauzi, M. Masrur, S. Mukodimah, Prediction of layer chicken disease using fuzzy analytical hierarcy process. Int. J. Eng. Technol. **7**(2.26), 90–94 (2018, June)
7. K. Shankar, Prediction of most risk factors in hepatitis disease using apriori algorithm. Res. J. Pharm. Biol. Chem. Sci. **8**(5), 477–484 (2017)
8. R.V.Q. Srikant, R. Agrawal, Mining association rules with item constraints, in *KDD*, vol. 97 (1997, August), pp. 67–73
9. K. Karthikeyan, R. Sunder, K. Shankar, S.K. Lakshmanaprabu, V. Vijayakumar, M. Elhoseny, G. Manogaran, Energy consumption analysis of virtual machine migration in cloud using hybrid swarm optimization (ABC–BA). J. Supercomput. (2018). https://doi.org/10.1007/s11227-018-2583-3
10. A.E. Hassanien, R.M. Rizk-Allah, M. Elhoseny, A hybrid crow search algorithm based on rough searching scheme for solving engineering optimization problems. J. Ambient. Intell. Hum. Ized Comput. (2018). https://doi.org/10.1007/s12652-018-0924-y
11. M. Elhoseny, A. Hosny, A.E. Hassanien, K. Muhammad, A.K. Sangaiah, Secure automated forensic investigation for sustainable critical infrastructures compliant with green comput-

ing requirements. IEEE Trans. Sustain. Comput. **PP**(99) (2017). https://doi.org/10.1109/tsusc.2017.2782737
12. M. Sajjad, M. Nasir, K. Muhammad, S. Khan, Z. Jan, A.K. Sangaiah, M. Elhoseny, S.W. Baik, Raspberry Pi assisted face recognition framework for enhanced law-enforcement services in smart cities, in *Future Generation Computer Systems* (Elsevier, 2018). https://doi.org/10.1016/j.future.2017.11.013
13. M.J. Zaki, Parallel and distributed association mining: a survey. IEEE Concurr. **7**(4), 14–25 (1999)
14. D. Ramesh, B. Vishnu Vardhan, Data mining techniques and applications to agricultural yield data. IJARCCE **2**(9), (2013, September)
15. I. Jagielska, C. Mattehews, T. Whitfort, An investigation into the application of neural networks, fuzzy logic, genetic algorithms, and rough sets to automated knowledge acquisition for classification problems. Neurocomputing **24**, 37–54 (1999)
16. K. Verheyen, D. Adriaens, M. Hermy, S. Deckers, High resolution continuous soil classification using morphological soil profile descriptions. Geoderma **101**, 31–48 (2001)
17. S. Veenadhari, Crop productivity mapping based on decision tree and Bayesian classification. Unpublished M.Tech Thesis submitted to Makhanlal Chaturvedi National University of Journalism and Communication, Bhopal, 2007
18. A. Chinchulunn, P. Xanthopoulos, V. Tomaino, P.M. Pardalos, Data mining techniques in agricultural and environmental sciences, Int. J. Agric. Environ. Inf. Syst. **1**(1), 26–40 (2010, January–June)
19. S. Veenadhari, B. Misra, C.D. Singh, Data mining techniques for predicting crop productivity—a review article. Int. J. Comput. Sci. Technol. (IJCST) **2**(1) (2011, March)
20. D. Shalvi, N. De Claris, Unsupervised neural network approach to medical data mining techniques, in *Proceedings of IEEE International Joint Conference on Neural Networks*, Alaska (1998, May), pp. 171–176
21. B. Rajagopalan, U. Lal, A K-nearest neighbor simulator for daily precipitation and other weather variable. Water Resour. **35**, 3089–3101 (1999)
22. A. Tellaeche, X.P. BurgosArtizzu, G. Pajares, A. Ribeiro, A vision-based hybrid classifier for weeds detection in precision agriculture through the Bayesian and Fuzzy k-Means paradigms, in *Innovations in Hybrid Intelligent Systems* (Springer, Berlin, Heidelberg, 2007), pp. 72–79
23. K. Somvanshi et al., Modeling and prediction of rainfall using artificial neural network and arima techniques. J. Ind. Geophys. Union **10**(2), 141–151 (2006)
24. A. Mucherino, P. Papajorgji, P. Pardalos, Data Mining in Agriculture, vol. 34 (Springer, 2009)
25. A. Urtubia, J.R. Pérez-Correa, A. Soto, P. Pszczolkowski, Using data mining techniques to predict industrial wine problem fermentations. Food Control **18**(12), 1512–1517 (2007)
26. X. Wu, X. Zhu, G.-Q. Wu, W. Ding, Data mining with Big Data. Trans. Knowl. Data Eng., **26**(1) (2014, January). 1041–4347/14, IEEE
27. S. Beniwal, J. Arora, Classification and feature selection techniques in data mining. Int. J. Eng. Res. Technol. (IJERT) **1**(6), 7 (2012). LiorRokach, OdedMaimon, “Clustering Methods”, Chap. 15
28. R. Xu, D. Wunsch, Survey of clustering algorithms. IEEE Trans. Neural Netw. **16**(3), 645–678 (2005)
29. U. Fayyad, G. Piatetsky-Shapiro, P. Smyth, From data mining to knowledge discovery in databases. AI Mag. **17**(3), 37 (1996)

WordNet Ontology-Based Web Page Personalization Using Weighted Clustering and OFFO Algorithm

N. Balakumar and A. Vaishnavi

Abstract With the development of information technology, personalization is recognized as one of the emerging technologies in the research field. Web page personalization process the user's query and retrieve the search results that correspond to their interest. In some cases, it is difficult to identify the desired result when the user having a different background on the same query. It is overcome by the proposed work; here the web page personalization is done through the query formulation and profiling by the WordNet ontology. Initially, the required data are collected from the web sources and are clustered using the presented Weighted Clustering (WC) algorithm. The WC clusters the web pages in corresponds to their domains and then, it is learned by the user learning module. With the help of four similarity measures, data similarity is evaluated between the generated word net and trained dataset. From that, the maximum similarity based data is achieved by the proposed algorithm called Oppositional based FireFly Optimization (OFFO). The results demonstrate that the WC-OFFO attains the precision of 89.16, recall of 78.09 and f-measure of 83.26 which is high compared to existing algorithms.

Keywords Web page personalization · Clustering · WordNet · User query · Similarity measures · OFFO algorithm

1 Introduction

With the gigantic measure of data accessible on the World Wide Web, some Internet users may confront data overflow issues, where there is an enormous number of facilitated reports on the web [1]. By convention, learning systems ought to give custom directions to students on finding the suitable learning materials [2], personalized and adjusted to the necessities, information, gifts and learning style of learners [3].

N. Balakumar (✉) · A. Vaishnavi
Department of Computer Applications, Pioneer College of Arts and Science, Coimbatore, India
e-mail: balakumar198392@gmail.com

A. Vaishnavi
e-mail: vaishmsc@gmail.com

M. Elhoseny and A. K. Singh (eds.), *Smart Network Inspired Paradigm and Approaches in IoT Applications*, https://doi.org/10.1007/978-981-13-8614-5_9

The keyword-based search engines can't fulfill the user needs at the time of his web search process [4]. Personalization of web search results can explain the specified issues by concentrating on the most important outcomes for the user inquiry combined with his inclinations recognized in his user profile [5]. Moreover, it diminishes the user endeavors at the time of the web search process. As of late, a few methodologies of web search personalization have been proposed [6]. Some of the effective methodologies depend on building a word net that indicates user interests and utilizing it in the web search process [7, 8]. Assorted references of a group of onlookers' forces content providers to generally utilize personalization innovations [9]. Web page personalization includes pleasing between people by finding their inclinations and utilizing these discovered inclinations to find the most applicable substance to every user [10].

In web page personalization system, alterations with respect to the searching content or perhaps the structure of an online web website are performed powerfully [11]. Thus a very much composed altered distance instruction system may have the resulting attributes [12]: (i) It will discover the students' investigation premium, get to propensities, learning introduction (ii) It will change the area and customize the students' interface powerfully predictable with the entrance log (iii) It will propose learning assets by examining the leaner's interest as well as learning process (iv) It will prescribe intrigued data (v) It will give recommendations to help him/her to manage teaching set up, teaching model and the way of teaching [13, 14]. Main parts of the Web personalization system are first, the classification and preprocessing of Web information, besides, [15] the extraction of relationships betweens and across various types of such learning, lastly, the assurance of the activities that should be advised by such a web personalization framework [16].

At the point, when the aim behind the search question isn't clear, the user will receive a vast number of results consequently. The user should be quick through an extensive rundown of results to discover the outcome that suits his data require [17]. Subsequently, a personalized learning system is required to give concerned data to the user by mining systems like clustering and optimization [18]. This paper proposes to combine the components of dynamic database systems and data mining methods to give an effective, adaptable, brilliant web personalization system.

Organization of the Paper: Analysis of Existing literature about web page personalization based on the user query and optimization techniques are presented in Sect. 2. The purpose of web page personalization is described in Sect. 3; Sect. 4 gives a detailed description of the proposed methodology along with the WC-OFFO techniques. Section 5 explains the results of the clustering and optimization work and the conclusion part is presented in Sect. 6 along with perspectives for this work.

2 Existing Literature—A Survey

In 2012 Moawad et al. [19] proposed another multi-specialist system based methodology for personalizing the web search results. With the help of the proposed system, build a user profile model from initial and basic information, and retain it via user feedback up-to-date user profile. During the searching process, the user query is optimized in two ways: by user profile preferences and WordNet ontology.

In 2014 Park [20] discussed the influence factors of Social networking sites (SNSs) during web page personalization. The affecting factors are: switching cost and satisfaction. This survey was evaluated with the samples of 677 SNS users from six universities in the US by using Structural Equation Modeling (SEM) technique. The results showed that, as expected, the personalization increases its switching cost as well as satisfaction, which results in the further use of SNSs.

A semantic invention of RESTful services according to the HATEOAS principle was illustrated by Ben Lamine et al. [21] author presented a clustering technique based on the similarity measures of the user query and generate an ontology to manage users and their services. The proposed technique has been implemented by a prototype to investigate the system effectiveness.

In 2018 Srinivasa Rao and Vasumathi [22] presented a web page personalization based on fuzzy-bat classification method. A user queried various search engines and merges the obtained results based on the links. Several calculations such as title, snippet, content, URL, link, and co-occurrence is calculated with the use of merged links. The evaluated values are assigned to the fuzzy-bat classifier to rank the list of links and then compare the proposed approach with existing fuzzy in terms of precision and response time.

In 2016 Chawla [23] presented an effective web page personalization based on clustered query sessions. The URLs are ranked related to the user query by utilizing genetic algorithm (GA). The performance of GA is analyzed on the data set of web query sessions in academics, entertainment, and sports. The results proved GA attains better personalization according to the user needed information.

Babu and Samuel [24] discussed the concept network to recognize users search inclinations. This network comprises the historical background of the user's previous search results. This approach recovers web pages that are identified with the user query. Genetic Algorithm (GA) was utilized when a user searches for something new and to compare the similarity of user's concept network with others. This helps to attain better search results in the field of the user query.

3 Importance of Web Page Personalization

Nowadays, the internet plays a very important role to satisfy the user needs, especially for different user queries at a simultaneous time. In some cases, the user may not satisfy by the retrieved information from the search engine, the reason is each user queried with a different intention. Personalization helps to find and presents right Information to the correct user at the correct time. Web Page Personalization model is to retrieve the best search results that meet the user's preferences using his up-to-date user profile, which is being built and updated regularly.

4 Methodology

The aim of web page personalization is to provide the personalized experience of different web pages based on known and relevant information of user query. In the proposed work, we collect the data from web page sources to personalize the data by the user preferences. At first, the collected data from Syskill and Webert Web Page Ratings are clustered (by using WC) according to their separate domains. After clustering, learn the data i.e. user needed information which is in the form of text, video or audio and stored with the help of learning content module. Next, to that, word net is generated which contains the keyword with its explanation. Based on the user query, we combine the trained dataset and word net generated database and find the similarity data by using different measures. The most minimum similarity valued data (user preferred page) is attained by the inspired optimization algorithm called OFFO from the inspiration of fireflies attractiveness. The diagrammatic representation of the proposed work is depicted in Fig. 1.

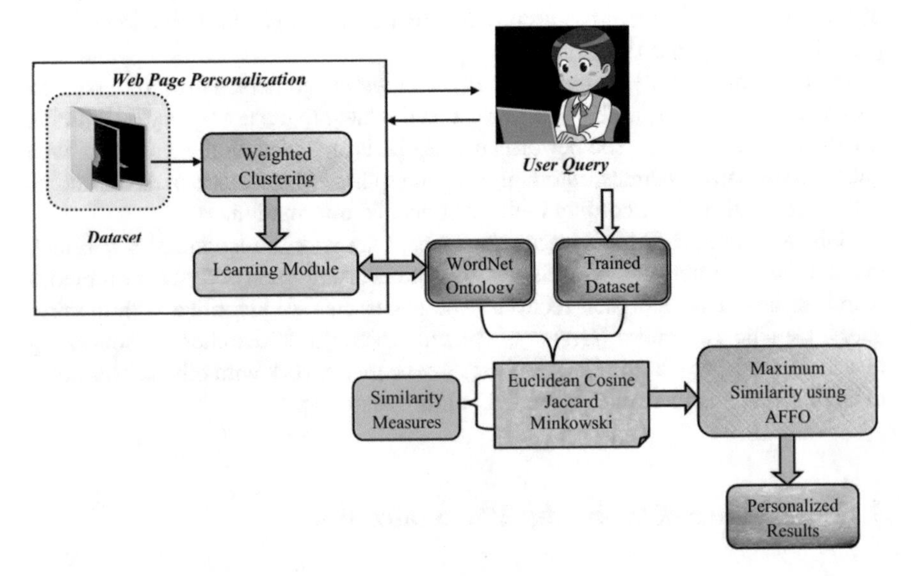

Fig. 1 The diagrammatic representation of the proposed work

4.1 Data Collection

The first step is to collect the data from the Syskill and Webert Web Page Ratings which contains the information related to web pages and their ratings of an individual user. The four separate subjects on the web pages are Bands-recording artists; Goats; Sheep; and Bio-Medical.

4.2 Clustering the Data Using WC Algorithm

The WC algorithm aims to cluster the subjects according to their domain by the user preferences. Clustering is done by the Cluster Head (CH) selection and in view of the every chosen CH; neighboring nodes i.e. having closer distances from CH are clustered at one. Similarly, three more CHs are selected and finally, clustered into four groups [25].

The procedure of WC Algorithm
Step 1: Find the neighbors of each node (k) by defining its degree d_k as

$$d_k = |N(k)| = \sum_{k' \in K, k; \neq k} \{distance(k, k') < node\, transmission\, range\} \quad (1)$$

Step 2: Evaluate the degree difference for every node k, $\Delta_k = |d_k - \delta|$ and then compute the sum of the distances with all its neighbors as:

$$Distance_k = \sum_{k' \in N(k)} \{distance(k, k')\} \quad (2)$$

Step 3: For every node, compute the running average of the speed till the present time t. This gives mobility $Mob_k = \frac{1}{T} \sum_{t=1}^{T} \sqrt{(A_t - A_{t-1})^2 + (B_t - B_{t-1})^2}$, where (A_t, B_t) and (A_{t-1}, B_{t-1}) are the coordinates of the node k at the time t and $t - 1$ respectively.
Step 4: Find the cumulative time C_k, during which a node k acts as a CH
Step 5: For each node, the combined weight is calculated as

$$CW_k = w_1 \Delta_k + w_2 Distance_k + w_3 Mob_k + w_4 C_k \quad (3)$$

where, $w_1, w_2, w_3\, and\, w_4$ are the weighting factors for the corresponding system parameters
Step 6: Cluster Head Selection: Select the node which has smallest combined weight as the CH. The neighbors of selected CH are no longer allowed to participate in the election procedure.

The above steps 2–6 are repeated until the rest of the nodes not yet selected as CH or assigned to a cluster. Based on this principle, we can group the data according to their domain.

4.3 Learning Module

Learning requires a set of positive models of user question (for example, web pages one is interested in) and negative examples (for example, web pages one is not interested in). In this paper, we learn a concept that recognizes pages by rating it by the user inclinations. A learner model needs to perpetually search the relevant data on the domain skills i.e. learn the data from the subjects bands, goats, sheep and bio-medical. It will responsible for creating as well as updating the user profile by web page rating.

4.4 WordNet

A WordNet is a tool for mapping the user understanding words, phrases and relationship between them or its synonyms term. This WordNet archive is useful in the applications of data mining as well as web mining.

The function of WordNet in web page personalization
When the user submits a query related to the bio-medical, bands, etc., the system finds the words (synonyms, phrases, etc.) related to a user query by using WordNet and makes a dataset. Search engine searches web for all words or phrases in synonym vector so gathers a complete set of web pages and delivers to web search personalization system. Web search personalization system selects relevant pages based on user interesting words in the user profile. The word net ontology is represented in Fig. 2.

4.5 Similarity Measures

To quantify or measure the similarity or consistency among the data items, the evaluation of distance metrics between the data items is essential. It is important to recognize, in what way the data are interrelated, how various data dissimilar or similar to one other and what measures are considered for their comparison. The measures used for finding the similarity phrases between trained and WordNet dataset are Minkowski, Euclidean, Chebyshev and Jaccard [26].

Initialized Parameter: U and V indicate the word or phrases from the trained and WordNet dataset respectively, the term i and j represents ith and jth data from the

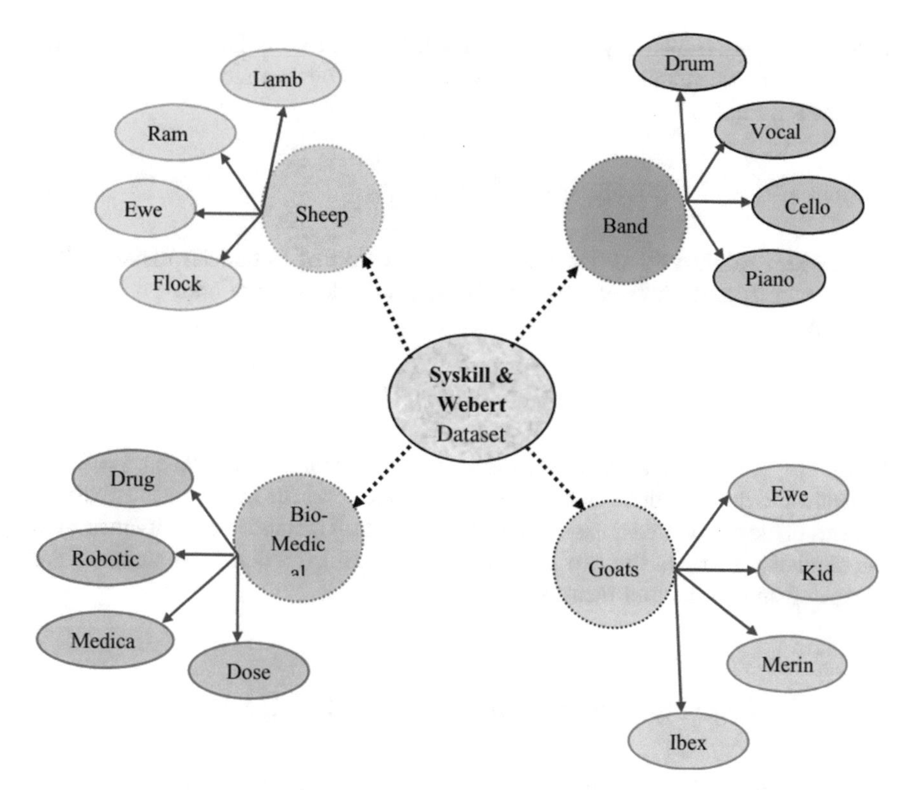

Fig. 2 WordNet based on proposed work

dataset. Based on this, the similarity measure between two words or phrases is calculated by using the four distance measure.

Euclidean: It is a standard distance measure for geometrical problems, it evaluates the difference between two data points. It is simply, the ordinary distance between two points.

$$Dis\tan ce_{AB} = \max_{k}\left|U_{ik} - V_{jk}\right| \tag{4}$$

Cosine distance: The function of cosine distance measure is to determine the cosine angle between the assigned two phrases or words i.e. user query. Here θ gives the angle between two vectors and it is evaluated by using the Eq. (5).

$$\theta = arc\ \cos\frac{U.V}{\|U\|\|V\|} \tag{5}$$

Jaccard: Jaccard is also a type of similarity distance measure. It measures the similarity of the two data items as the intersection divided by the union of the data items as shown in Eq. (6).

$$J(U, V) = \frac{|U \cap V|}{|U \cup V|} \tag{6}$$

Minkowski: Minkowski distance is a generalization of Euclidean distance. The Minkowski distance between two variables (similarity of two phrases) U and V is defined as

$$Distance_{UV} = \left(\sum\nolimits_{k=1}^{d} |U_{ik} - V_{jk}|^{\frac{1}{a}}\right)^{a} \tag{7}$$

From Eq. (7), note that when a = 1, it is equivalent to Manhattan distance; a = 2, it is equivalent to the Euclidean distance. Where d symbolizes the number of attributes or phrases in the considered dataset. Using the above four distance similarity measures, the similarity between the two intervals of the considered dataset is analyzed and compared among the four measures.

4.6 Web Page Personalization Using OFFO Algorithm

In a standard point of view, the process of optimization can be described to find the best solution of the function from the system within constraints [27–30]. According to the user query, the data is accessed to retrieve the related preferences with the highest priority. Here, the highest priority one is attained by the maximum distance similarity measure which is chosen by using the OFFO algorithm.

$$\text{Objective function } OF = \max(similarity\, between\, data) \tag{8}$$

FireFly Optimization (FFO): FFO algorithm is a metaheuristic one which is inspired by the flashing behavior of fireflies [31, 32]. In OFFO, the initialized parameters are improved by simultaneously checking the opposite solution and the fitter one can be chosen as an initial solution.

The behavior of Fireflies: Generally, fireflies are unisexual and are attracted to one other. However, the attractiveness decreased when the distance between the two fireflies increased and vice versa. If the brightness of both fireflies is the same, it will move randomly. The new generation or the updating process of fireflies is based on the characteristics of its random walk and attraction. The procedure involved in OFFO algorithm is explained as follows:

OFFO Initialization: Initialize the number of fireflies population f (here, the similarity measures of the data are initialized as the population)

$$F = F_i = \{F_1, F_2, F_3, \ldots F_n\} \tag{9}$$

Oppositional Process: Let $F \in (m, n)$ is a real number. By applying the opposite point definition, it can be written as

$$\tilde{F}_j = m_j + n_j - F_j \tag{10}$$

Fireflies Attractiveness Behavior: The attractiveness (brightness), B_{ness} of firefly 1 on the Firefly 2 is based on the degree of the brightness of the firefly 1 and the distance (d) between the Firefly 1 and 2. The brightness B_{ness} of a firefly is selected by its fitness value or objective function OF (maximum similarity).

$$B_{ness} = OF(F_i) \tag{11}$$

The expansion of Eq. (11) is: Suppose there are n numbers of fireflies (data), and $F_{(i=1,2\ldots n)}i$ corresponds to the position for Firefly. The function of attractiveness is described as

$$\eta(d) = \eta_0 e^{-\lambda d^2} \tag{12}$$

where the term η indicates the attractiveness value (based on the distance between two fireflies) and the λ represents the light absorption coefficient.

The movement towards Attractive Firefly: The movement of a Firefly 1 (less bright) at position F1 moving towards a brighter firefly 2 at position F2 and the Firefly 1's updated position is calculated by Eq. (13).

$$F_i(t+1) = F_i(t) + \eta_0 e^{-\lambda d^2}(F_1 - F_2) + \alpha\left(rand - \frac{1}{2}\right) \tag{13}$$

where $\eta_0 e^{-\lambda d^2}(F_1 - F_2)$ is due to the attraction of firefly F2 and the term $\alpha\left(rand - \frac{1}{2}\right)$ symbolizes the randomization parameter and rand is a random number generated uniformly in the range of 0 and 1. In the same way, using this OFFO algorithm, the data (with minimum similarity) is updated by the data (with maximum similarity) from each domain.

Termination Condition: The OFFO algorithm compares the similarities of the new data position with old one. If the new position produces higher similarity value, the data is moved to the new position; otherwise, it will remain in the current position. The termination criterion of the OFFO is based on a predefined number of iterations or fitness value or objective function. The flowchart of OFFO is depicted in Fig. 3.

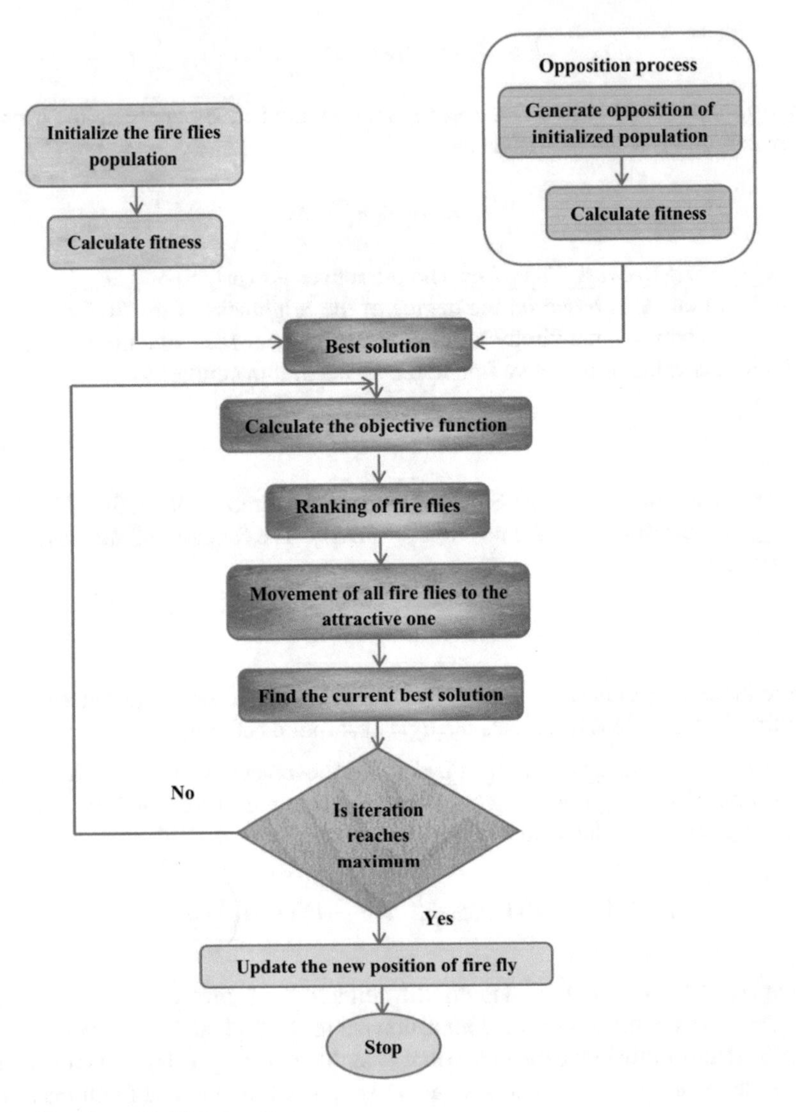

Fig. 3 Flowchart of OFFO algorithm

4.7 *Personalization of Maximum Similarity Data*

The Query Optimizer processes the user query to get the similarity of each word or phrases within the user query using the OFFO algorithm. According to the user query, the web page is personalized by accessing the WordNet ontology with reference to the related words. By using this OFFO algorithm, the user required data is personalized with maximum accuracy.

5 Result and Discussion

The proposed methodology is implemented using the working platform of MATLAB 2016a with the system configuration, i5 processors with 4 GB RAM. In the result analysis section, the performances of the proposed WC with OFFO algorithm is discussed and compared with existing techniques. The performance measures such as precision, recall, and f-measure are assessed and compared with the actual result.

5.1 Database Description

This database contains the HTML source of web pages plus the ratings of a single user on these web pages. The web pages are on four separate subjects (Bands- recording artists; Goats; Sheep; and BioMedical).

5.2 Performance Measures

Precision: Precision gives information about the effectiveness of the proposed system. Precision is defined as the ratio of a number of relevant pages retrieved to the number of pages retrieved [33–32].

$$P = \frac{n(\mathrm{Ret}_{\mathrm{Rel}})}{n(\mathrm{Ret})} \tag{14}$$

where,

$n(\mathrm{Ret}_{\mathrm{Rel}})$ Number of relevant pages retrieved
$n(\mathrm{Ret})$ Number of pages retrieved.

Recall: Recall gives information about the accuracy of the proposed system. The recall is defined as the ratio of a number of relevant pages retrieved to the total number of relevant pages in the database.

$$R = \frac{n(\mathrm{Ret}_{\mathrm{Rel}})}{n(Db_{\mathrm{Rel}})} \tag{15}$$

where Db_{Rel} is the total number of relevant pages in the database.

F-measure: F-measure is a measure of a harmonic mean of both precision and recall. It is described as in Eq. (16).

$$F - measure = 2.\frac{precision..recall}{precision + recall} \tag{16}$$

Table 1 Accuracy of similarity measures

Clustered subject	Cluster no.	The accuracy of similarity measures			
		Euclidean	Cosine	Jaccard	Minkowski
Band	3	76.25	76.52	77.11	79.34
	4	77.82	77.91	78.27	81.16
	5	80.35	81.06	81.63	83.22
	6	79.64	79.68	80.29	82.19
Goats	3	75.23	74.12	75.24	74.25
	4	75.68	78.68	80.14	78.25
	5	78.12	81.47	78.56	77.68
	6	78.23	82.12	76.35	74.24
Sheep	3	74.36	74.44	80.24	80.01
	4	76.12	76.84	80.12	79.64
	5	80.12	77.41	81.47	75.14
	6	78.69	73.54	79.56	74.44
Bio-medical	3	76.45	75.08	77.87	77.78
	4	77.01	79.68	80.24	78.64
	5	81.23	77.23	80.29	78.14
	6	79.28	74.36	78.56	76.32

Table 1 explains the accuracy of similarity measures attained between the data for the considered datasets i.e. Syskill and Webert Web Page Ratings. Based on the data similarity, web pages are clustered as band, goats, sheep, and bio-medical by using weighted clustering model. For the constrained dataset, the accuracy of Minkowski is 83.22, Jaccard is 81.63, Cosine 81.06 and Euclidian is 80.35 at the cluster number 5. Similarly, these four measures are calculated for the cluster 3, 4 and 6. On comparing the four distances based similarity measures, Minkowski distance measure attains high accuracy for all clusters.

Performance Analysis

Figure 4 illustrates the precision analysis of web page personalization from the Syskill and Webert Web Page Rating dataset by using different algorithms like WC with OFFO, WC-FFO, and GA. By the utilization of these algorithms, the optimal or maximum similarity measures between user query data among the four subjects such as band, goats, sheep and bio-medical are calculated. On comparing the algorithms, the proposed model WC with OFFO attains high precision for all the clusters.

The performance measures such as recall and F-measure of WC with OFFO, WC-FFO, and GA are analyzed and illustrated in Figs. 5 and 6 respectively. By using these algorithms, the optimal or maximum similarity measures between user query data among the four subjects such as band, goats, sheep and bio-medical are calculated. The highest recall of 78.09 is attained in the 5th cluster for the proposed WC-OFFO algorithm and the maximum f-measure of 83.25 is achieved in the 5th cluster. By

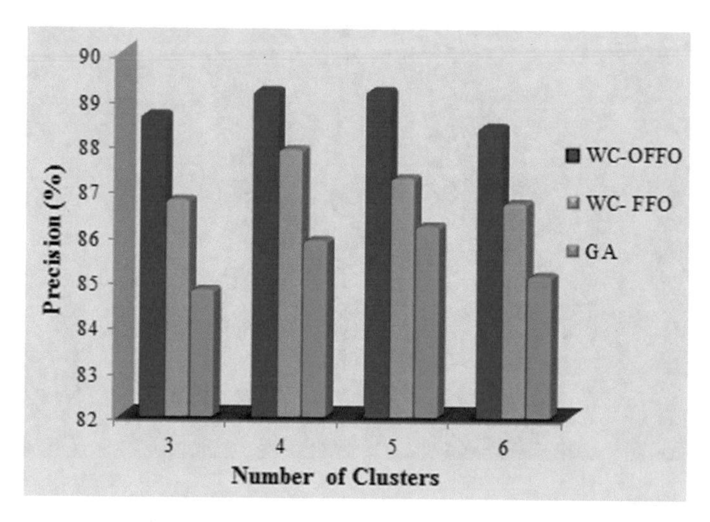

Fig. 4 Precision

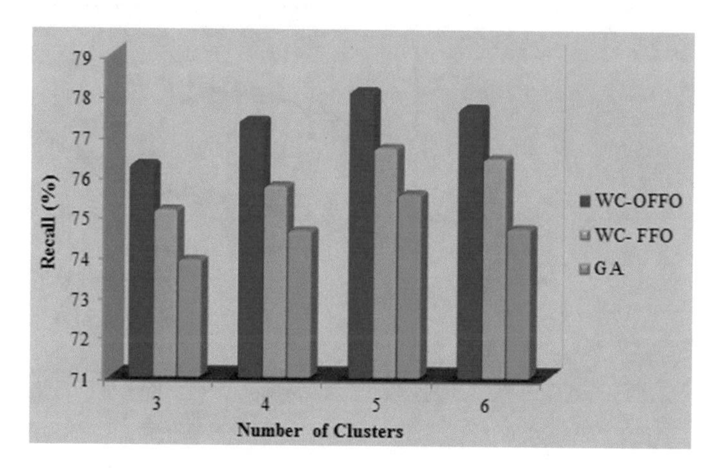

Fig. 5 Recall

analyzing these measures (recall and f-measure), the efficiency of optimization algorithms is evaluated.

Comparative Analysis of Web page Personalization

The comparative analysis of personalized and google results for the user searched query is depicted in Fig. 7. In this work, user personalized the words or phrases related to four different subjects such as band, goats, sheep and bio-medical. On considering web page personalization, their data relevancy is high compared to google searched results for all the four subjects.

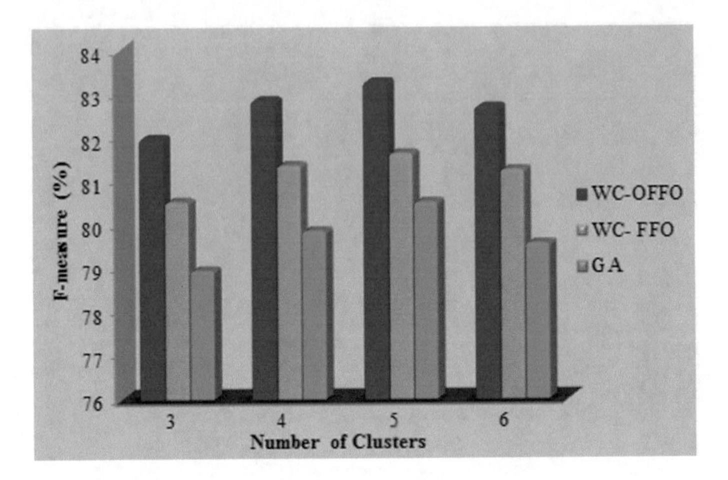

Fig. 6 F-measure

Fig. 7 Comparative analysis

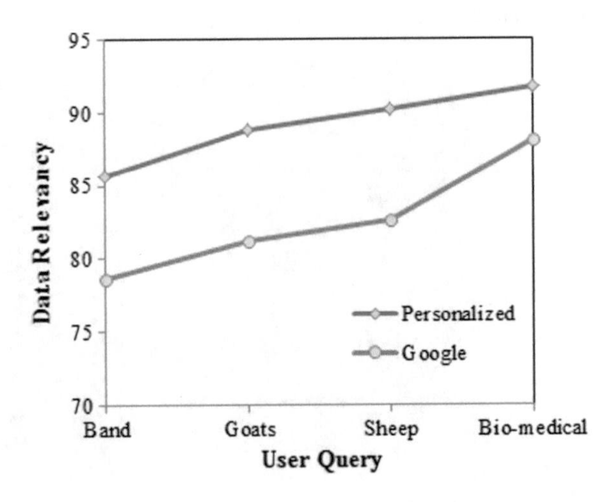

Execution Time Analysis

Figure 8 shows the time taken for personalizing the maximum similarity data based on a user query (the query is related to four subjects like band, goats, sheep and bio-medical). The computational processing time is analyzed for three algorithms that are: Genetic Algorithm (GA), FFO and the proposed OFFO. For cluster 3, the GA takes 223,154 s to choose the maximum similarity measure data, FFO takes 212,154 s and the proposed Opposition based FFO requires less time i.e. 192,642 s compared to other algorithms. Similarly, the time analysis is evaluated and compared for the cluster 4, 5, and 6.

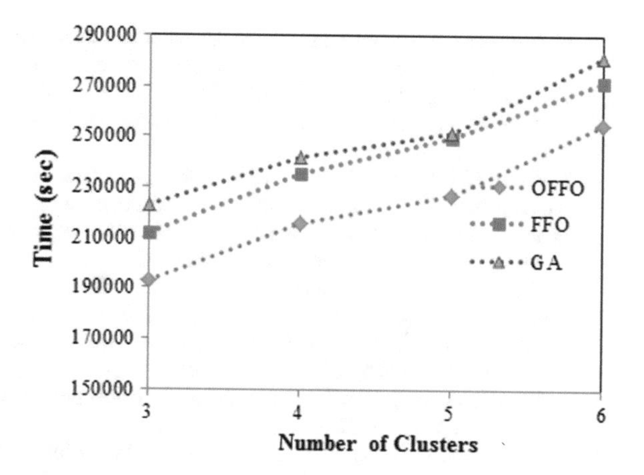

Fig. 8 Time analysis

6 Conclusion

The paper discussed web page personalization for retrieving the search results of the user query. In this paper, the desired result of user search query is retrieved even if there are an infinite number of related words on the same query by using WordNet ontology. Initially, the data collected from Syskill and Webert Web Page Rating dataset are clustered according to their corresponding subjects by the proposed WC model. The data with maximum similarity measures were clustered by choosing the Cluster Head. Web page personalization is done by comparing the similarity measures of trained and WordNet and then the maximum similarity between data is attained by the proposed optimization algorithm i.e. OFFO. The OFFO algorithm compares the similarities of the new data position with the old one. If the new position produces higher similarity value, the data is moved to the new position; otherwise, it will remain in the current position. The results demonstrated that the WC-OFFO model improves the personalization efficiency by achieving high-performance measures such as precision of 89.16, recall of 78.09 and f-measure of 83.26 compared to existing algorithms. For future examinations, web page personalization efficiency is improved by other innovative clustering and hybrid optimization model.

References

1. P.Y.K. Chau, S.Y. Ho, K.K.W. Ho, Y. Yao, Examining the effects of malfunctioning personalized services on online users' distrust and behaviors. Decis. Support Syst. **56**, 180–191 (2013)
2. K. Makvana, P. Shah, P. Shah, A novel approach to personalize web search through user profiling and query reformulation, in *2014 International Conference on Data Mining and Intelligent Computing (ICDMIC)* (2014)
3. K.S. Kuppusamy, G. Aghila, CaSePer: an efficient model for personalized web page change detection based on segmentation. J. King Saud Univ. Comput. Inf. Sci. **26**(1), 19–27 (2014)

4. G. Bordogna, A. Campi, G. Psaila, S. Ronchi, Disambiguated query suggestions and personalized content-similarity and novelty ranking of clustered results to optimize web searches. Inf. Process. Manag. **48**(3), 419–437 (2012)
5. N.M. Markovich, Analysis of clusters in network graphs for personalized web search. IFAC-PapersOnLine **50**(1), 5178–5183 (2017)
6. S. Knerr, K.J. Wernli, K. Leppig, K. Ehrlich, A.L. Graham, D. Farrell, S.C. O'Neill, A web-based personalized risk communication and decision-making tool for women with dense breasts: design and methods of a randomized controlled trial within an integrated health care system. Contemp. Clin. Trials **56**, 25–33 (2017)
7. F. Grandi, Dynamic class hierarchy management for multi-version ontology-based personalization. J. Comput. Syst. Sci. **82**(1), 69–90 (2016)
8. X. Peng, Z. Niu, S. Huang, Y. Zhao, Personalized web search using clickthrough data and web page rating. JCP **7**(10), 2578–2584 (2012)
9. S. Vanitha, A personalized web search based on user profile and user clicks. Int. J. Latest Res. Sci. Technol. **2**(5), 78–82 (2013)
10. V. Viswanathan, K. Ilango, Ranking semantic relationships between two entities using personalization in context specification. Inf. Sci. **207**, 35–49 (2012)
11. L. Wanner, M. Rospocher, S. Vrochidis, L. Johansson, N. Bouayad-Agha, G. Casamayor, L. Serafini, Ontology-centered environmental information delivery for personalized decision support. Expert. Syst. Appl. **42**(12), 5032–5046 (2015)
12. A. Hawalah, M. Fasli, Utilizing contextual ontological user profiles for personalized recommendations. Expert Syst. Appl. **41**(10), 4777–4797 (2014)
13. C. Liang, User profile for personalized web search, in *2011 Eighth International Conference on Fuzzy Systems and Knowledge Discovery (FSKD)* (2011)
14. T.T. Dao, T.N. Hoang, X.H. Ta, M.C. Ho Ba Tho, Knowledge-based personalized search engine for the web-based human musculoskeletal system resources (HMSR) in biomechanics. J. Biomed. Inform. **46**(1), 160–173 (2013)
15. D. Yoo, Hybrid query processing for personalized information retrieval on the semantic web. Knowl.-Based Syst. **27**, 211–218 (2012)
16. M. Sah, V. Wade, Personalized concept-based search on the linked open data. SSRN Electron. J. (2016)
17. Y. Du, Y. Hai, Semantic ranking of web pages based on formal concept analysis. J. Syst. Softw. **86**(1), 187–197 (2013)
18. Y. Guan, D. Zhao, A. Zeng, M.-S. Shang, Preference of online users and personalized recommendations. Phys. A **392**(16), 3417–3423 (2013)
19. I.F. Moawad, H. Talha, E. Hosny, M. Hashim, Agent-based web search personalization approach using dynamic user profile. Egypt. Inform. J. **13**(3), 191–198 (2012)
20. J.-H. Park, The effects of personalization on user continuance in social networking sites. Inf. Process. Manag. **50**(3), 462–475 (2014)
21. S.B.A. Ben Lamine, H. Baazaoui Zghal, M. Mrissa, C. Ghedira Guegan, An ontology-based approach for personalized RESTful Web service discovery. Procedia Comput. Sci. **112**, 2127–2136 (2017)
22. P. Srinivasa Rao, D. Vasumathi, Utilization of co-occurrence pattern mining with optimal fuzzy classifier for web page personalization. J. Intell. Syst. **27**(2), 249–262 (2018)
23. S. Chawla, A novel approach of cluster based optimal ranking of clicked URLs using genetic algorithm for effective personalized web search. Appl. Soft Comput. **46**, 90–103 (2016)
24. K.R.R. Babu, P. Samuel, Concept networks for personalized web search using genetic algorithm. Procedia Comput. Sci. **46**, 566–573 (2015)
25. M. Chatterjee, S.K. Das, D. Turgut, WCA: a weighted clustering algorithm for mobile ad hoc networks. Clust. Comput. **5**(2), 193–204 (2002)
26. S. Bodkhe, M. Padole, An efficient methodology for clustering uncertain data based on similarity measure. J. Comput. Eng. (IOSR-JCE) **18**(4), 12–16 (2016)

27. S.K. Lakshmanaprabu, K. Shankar, D. Gupta, A. Khanna, J.J.P.C. Rodrigues, P.R. Pinheiro, V.H.C. de Albuquerque, Ranking analysis for online customer reviews of products using opinion mining with clustering. Complexity **2018**, Article ID 3569351, 9 (2018). https://doi.org/10.1155/2018/3569351
28. A.R.R. Hosseinabadi, J. Vahidi, B. Saemi, A.K. Sangaiah, M. Elhoseny, Extended genetic algorithm for solving open-shop scheduling problem. Soft Comput. (Springer) (2018). https://doi.org/10.1007/s00500-018-3177-y
29. S.K. Lakshmanaprabu, K. Shankar, A. Khanna, D. Gupta, J.J. Rodrigues, P.R. Pinheiro, V.H.C. De Albuquerque, Effective features to classify big data using social internet of things. IEEE Access **6**, 24196–24204 (2018)
30. K. Shankar, S.K. Lakshmanaprabu, D. Gupta, A. Maseleno, V.H.C. de Albuquerque, Optimal feature-based multi-kernel SVM approach for thyroid disease classification. J. Supercomput. (2018). https://doi.org/10.1007/s11227-018-2469-4
31. N.F. Johari, A.M. Zain, M.H. Noorfa, A. Udin, Firefly algorithm for optimization problem, in *Applied Mechanics and Materials*, vol. 421 (Trans Tech Publications, 2013), pp. 512–517
32. K. Shankar, M. Elhoseny, S.K. Lakshmanaprabu, M. Ilayaraja, R.M. Vidhyavathi, M. Alkhambashi, Optimal feature level fusion based ANFIS classifier for brain MRI image classification. Concurr. Comput. Pract Exper. e4887 (2018). https://doi.org/10.1002/cpe.4887
33. N. Metawaa, M. Kabir Hassana, M. Elhoseny, Genetic algorithm based model for optimizing bank lending decisions. Expert. Syst. Appl. (Elsevier) **80**, 75–82 (2017). https://doi.org/10.1016/j.eswa.2017.03.021
34. K. Karthikeyan, R. Sunder, K. Shankar, S.K. Lakshmanaprabu, V. Vijayakumar, M. Elhoseny, G. Manogaran, Energy consumption analysis of Virtual Machine migration in cloud using hybrid swarm optimization (ABC–BA). J. Supercomput. (2018). https://doi.org/10.1007/s11227-018-2583-3
35. A.E. Hassanien, R.M. Rizk-Allah, M. Elhoseny, A hybrid crow search algorithm based on rough searching scheme for solving engineering optimization problems. J. Ambient. Intell. Hum. Comput. (2018). https://doi.org/10.1007/s12652-018-0924-y

Mobility Condition to Study Performance of MANET Routing Protocols

Hind Ziani, Nourddine Enneya, Jihane Alami Chentoufi and Jalal Laassiri

Abstract The performance of a Mobile Ad hoc Network (MANET) is closely related to the capability of routing protocols to adapt themselves to unpredictable changes of topology network and link status. In simulations, performance studies of routing protocols depend on the chosen mobility model. Consequently, the comparison of the obtained performance results becomes more difficult even it runs on the same simulation environment. To solve this problem it is necessary add a mobility condition, independently, of the used mobility models. In this paper, we define this mobility condition and show how it makes performance comparison of routing protocols judicious and easier.

Keywords Mobility · Node mobility · Network performance · QoS · Routing protocols · Ad hoc networks · MANETs · MANET routing protocols

1 Introduction

A mobile ad hoc network (MANET) [1] is a collection of mobile nodes that cooperatively communicate with each other without any pre-established infrastructure such as a centralized access point. These nodes may be computers or devices such as laptops, PDAs, mobile phones and pocket PCs which have in common a wireless connectivity. The idea of forming a network with no infrastructure originates from DARPA (Defence Advanced Research Projects Agency) packet radio network's

---

H. Ziani · N. Enneya (✉) · J. A. Chentoufi · J. Laassiri
Informatics, Systems and Optimization Laboratory, Faculty of Sciences,
University of Ibn Tofail, 133 Kenitra, Morocco
e-mail: enneya@uit.ac.ma

H. Ziani
e-mail: hind.ziani@uit.ac.ma

J. A. Chentoufi
e-mail: j.alami@uit.ac.ma

J. Laassiri
e-mail: laassiri@uit.ac.ma

M. Elhoseny and A. K. Singh (eds.), *Smart Network Inspired Paradigm and Approaches in IoT Applications*, https://doi.org/10.1007/978-981-13-8614-5_10

days [2]. Due to the fact that nodes change their physical location by moving around, the network topology may unpredictably change, which causes changes in link status between each node and its neighbors. Thus, nodes which join and/or leave the communication range of a given node in the network will certainly change its relationship with its neighbors by detecting new link breakages and/or link additions. This can produce a large number of updates in the routing table of each node in MANET. Furthermore this topology change makes an overhead traffic in the process of route maintenance guaranteed by the implemented routing protocols in MANETs. So, the performance of a MANET is closely related to the capability of the routing protocol to adapt itself to topology changes and the link status [3, 4]. There are three main categories of MANET routing protocols: Proactive (table-driven), Reactive (on-demand) and Hybrid. Proactive protocols build their routing tables continuously by broadcasting periodic routing updates through the network; reactive protocols build their routing tables on demand and have no prior knowledge of the route they will take to get to a particular node. Hybrid protocols create reactive routing zones which are interconnected by proactive routing links and usually adapt their routing strategy to the amount of mobility in the network.

Simulating routing protocols in a network simulator have several benefits over real-world testing. Simulations allow researchers to evaluate the performance of routing protocols in a wide range of scenarios at no cost. For this reason, several mobility models were developed to simulate the pattern movement of nodes that will be followed during the simulation [5].

To evaluate MANET protocols, it is not suitable to use only one mobility model. Various models that span across all different mobility characteristics are needed. When evaluating a single protocol, this protocol is run on various models to see how its performance changes on different models. It is found that the performance of a specific protocol varies if underlying mobility models are different. When evaluating a group of protocols, these protocols are run on a single mobility model to see how these protocols behave with this modeled motion. It has been noticed that the behavior of protocols depends on mobility models used. So, it is very difficult to draw a conclusion on the performance of routing protocols from the obtained results depending on the chosen mobility models. Thus, to make this comparison possible, it's necessary to make simulations in the same *mobility condition*. To formulate this mobility condition we used a quantitative metric of mobility [4] that is independent of any used mobility model. This metric measures mobility of network when each simulation takes end.

In this paper, we present a unified quantitative metric of the mobility related to the change of link status. This mobility metric is calculated at regular discrete time intervals. Moreover, it is characterized by two main properties. First, logical because it has a strong linear relationship with the change of link status for a wide range type of scenarios. Second, lightweight because it is light in calculation and recalculation: it is easy to calculate it at regular intervals (during the exchange of HELLO messages) with less consumption in memory and CPU resources, which is suitable for real MANET.

The paper is organized as follows. In Sect. 2, mobility models are given and discussed. Section 3 presents our proposed metric of mobility. Section 4 firstly describes the main performance metrics that can evaluate a MANET, secondly it presents simulations and discusses the obtained results . The last section concludes the paper and gives an idea about our future works.

2 Mobility Models Overview

Mobility models are important because they determine the behavior of mobile nodes (MN) on stage [5, 6]. They can be classified into two types: those based on traces (logs of actual movements) and the synthetic (emulate reality by mathematical equations). Some authors classify mobility models into three groups [6]: models based on strokes (work with real mobility), models based on topology restrictions (real scenario simulations) and statistical models (study from randomness). Ad hoc networks do not work yet on models based on traces on the network characteristics. However, it is expected that study will expand in future on the application of these models [5, 6]. Therefore, models of synthetic mobility are used together with simulated scenarios. In order to prove this form of controlled mobility, certain parameters are used, which allow to obtain quantifiable date and thus to transform them into useful informaties. The synthetic models are classified according to their relationship with the representation of human mobility: synthetic mobility models unrealistic, for example: random models [6–8] (Random Walk Mobility Model, Random Waypoint Mobility Model), temporal dependency models [7, 9] (Boundless Simulation Area Mobility Model, Gauss-Markov Mobility Model, Smooth Random Mobility Model) and realistic synthetic mobility models such as: spatial dependence models [9, 10] (Reference Point Group Mobility, Column Mobility Model, Pursue Mobility Model, Nomadic Mobility Model) Geographic Restriction Models [10, 11] (Pathway Models, Obstacle Models, Human Obstacle Mobility Model).

3 Used Mobility Metric

3.1 Mobility Metrics Overview

Different mobility models lead to different mobility patterns. But models themselves do not give clear images how mobility patterns are different with each others. We need some mobility metrics to describe these mobility patterns. Efforts to find appropriate mobility metrics have begun only recently. We classify mobility metrics in two categories: direct mobility metrics and derived mobility metrics. The direct mobility metrics, like host speed or relative speed, are measurements with a clear physical meaning. The derived mobility metrics, like graph connectivity, are measurements derived from physical observations through mathematical modelling. The direct mobility metrics measure host motion directly, e.g., average host speed or minimum/maximum speed. For RandomWayPoint Model, pause time is also used to reflect node mobility [3, 12], namely, the longer the pause time is, the smaller the mobility. Other metrics belonging to this category include average relative speed [13], average degree of spatial dependence and temporal dependence [14]. Average relative speed [13] is defined based on relative speed of all pairs of hosts in the network. Attempts have been made in [14] to characterize the temporal dependence of the movement of an individual host and the spatial dependence between different hosts. The temporal dependence indicates how an individual host changes its velocity over time, or say, whether its current velocity is dependent on the previous velocity. Average degree of temporal dependence is proposed to capture temporal dependence. It is an average over the temporal dependence of all the hosts. For each host, the degree of temporal dependence is defined as the product of relative direction and relative speed (relative to itself) at two different times. The spatial dependence indicates whether a hosts movement is correlated with other hosts. The average degree of spatial dependence is the average of degree of spatial dependence over all host pairs. The direct mobility metrics has been used to measure different mobility models. For example, average degree of spatial dependence differentiates different mobility models successfully [14]; average relative speed varies almost linearly with link change rate (see next subsection) under Random WayPoint Model. However, some metrics can not accurately capture different characteristics of the models. For example, average degree of temporal dependence fails to differentiate different mobility models [15]. Average or minimum/maximum speed has been used widely. Although it indicates the degree of mobility; it fails to reflect relative motions between hosts. The metric “pause time” is model dependent: it can only be used in the Random WayPoint Model. More important, direct mobility metrics often do not directly reflect topology changes, while the latter is believed to be more influential to network performance. Take the RandomWalk Model for example; high mobility speed doesn’t necessarily generate large geographic movement [16] to cause dramatic topology changes. Mobility models impact the connectivity graph which in turn influences the protocol performance. It is thus helpful to study metrics that capture the properties of connectivity graph. The category of derived mobility metrics include metrics derived from graph theoretic models as well as other mathematical models. Metrics derived from graph-theoretic models include link change rate [16, 17], link duration [14, 18] and path duration [14]. The mobility measure metric proposed in [4] is derived from probabilistic models. Papers [16, 17] proposed link change rate as an indicator of topology change. If a link between two hosts is established/severed due to host movement we consider the state of the link between them up/down. Link change rate is the total number of link up/downs in unit time. average link duration [14, 18] is defined as the average of link durations over the host pairs that are within each others transmission range. The link duration is the time interval during which two hosts are within each others transmission range. Average path duration [14] averages the durations of all the paths linking every source destination pairs. Path duration is the time interval during which all links on a path (from a source to a destination) exist. The average path duration is related to the path length (hop count) h, average relative speed V and transmission range R. Mobility measure [4] is derived from the average relative speed. It is based on the observation that relative speeds do not make much sense for two nodes that are far away, but make much sense for two nodes that are near the transmission range of each other. A relation of remoteness between two nodes is defined as a function of the distance between two nodes; it increases from 0 to 1 monotonically. The derivative of remoteness is 0 at distance 0, increases as the distance increases, reaches its maxima at the communication boundary; then decreases as distance increases further, and approaches 0 as the distance approaches infinite. The mobility measure is defined as the average of the derivative of remoteness over all node pairs. Evaluations have been performed to investigate how the derived mobility metrics are related to direct mobility metrics, how well the derived metrics can differentiate different mobility models, and how well the metrics can quantify routing performance. Results from [16] show that the link change rate increases as average host speed increases. But results also show that, for different mobility models, differences in link change rate are small, which means link change rate can not differentiate different mobility models effectively. Moreover, [18] pointed out that the drawback of the link change rate is that it only counts the number of link changes without taking into account the duration of a link which heavily influences protocol performance. To this extent, [18] argues that average link duration is a good metric that not only quantifies host movements but also indicates protocol performance accurately. Under Random WayPoint Model, when average link duration increases, throughput increases and end-to-end delay and protocol overhead decreases consistently. For average path duration, it is found that at a high speed, path duration always shows exponential distribution no matter what mobility model is in use; it is also found that there exists a linear relationship between the reciprocal of the average path duration and the routing protocol performance in terms of throughput and routing overhead [14]. For the mobility measure defined on remoteness, simulations show that it has a consistent linear relationship with the link change rate for various mobility models [4].

3.2 Used Mobility Metric

In this section, we define our unified metric of mobility that is independent of any used mobility model, and related to the change of link status in the network. This mobility metric is calculated at regular discrete time intervals. Moreover, it is characterized by two main properties. First, logical because it has a strong linear relationship with the change of link status for a wide range type of scenarios. Second, light-weight because it is light in calculation: it is directly based on the data structure of neighbors used by each node in the ad-hoc network, and dont require any supplement process.

Based on the number of nodes leaving and/or joining the communication range of a given node in the MANET, we define two mobility measures in mobile ad-hoc networks. The evaluation of these two metrics can be made at regular time intervals as follows:

$$M_i^{\lambda}(t) = \lambda \frac{NodesOut(t)}{Nodes(t - \Delta t)} + (1 - \lambda)\frac{NodesIn(t)}{Nodes(t)} \tag{1}$$

After the estimation of the node mobility, we define the *wireless network mobility* at regular time intervals as follows:

$$Mob_{\lambda}(t) = \frac{1}{N}\sum_{i=0}^{N-1} M_i^{\lambda}(t) \tag{2}$$

where N is the number of nodes in the network. In addition, for a network in steady state, we can use the time average of the mobility measure as follows:

$$M_{\lambda} = \frac{\Delta t}{T}\sum_{k} Mob_{\lambda}(t) \tag{3}$$

where $k \in \{\Delta t, 2 * \Delta t, ..., [T/\Delta t] * \Delta t\}$. The [.] is the integer part and T is the total simulation time.

After presentation of the mobility properties of the proposed four mobility models related to the four mobility spaces including spatial and temporal dependence, relative speed, and geographic restrictions, we then focus on answering the following questions:

1. *Why dont, we get the same performance results through different mobility models of a MANET protocol simulated in the same network parameters?*
2. *What conditions must be considered to compare performance MANET protocols in different mobility models?*

4 Simulation and Results

To answer these questions, we have chosen to evaluate performance of the Optimized Link State Routing (OLSR) protocol [19], one of the proactive MANET routing protocols. Precisely, we have used the Network Simulator NS-2 [20] and the compatible OLSR implementation [21] to measure the three main metrics of performance cited in previous chapter (Fig. 1).

Table 1 Simulation parameters used in network simulator NS2 for OLSR performance evaluation

Simulation settings	
Nodes number	50 nodes
Topology area	1000 m × 1000 m
Transmission range	100 m
Traffic type	Canstat Bit Rate (CBR)
Connection rate	4 pkts/s
Packet size	512 bytes
Connections number	10
Simulation time	500 s

4.1 Simulation Environment

This OLSR evaluation was made for two mobility models (RWP and RGM models) in the worst case by supposing that the maximum speed of nodes is equal to Vmax = 40 m/s and the pause time is equal to 0 in case of Random Waypoint model. Moreover, simulations are considered in the same MANET environment as illustrated in the Table 1.

4.2 Performance Metrics

We have considered the more important metrics for analyzing and evaluating performance of MANET routing protocols. These considered metrics are:

- *Normalized Routing Overhead (NRL)*: It represents the ratio of the control packets number propagated by every node in the network to the data packets number received by the destination nodes. This metric reflects the efficiency of the implemented routing protocols in the network.
- *Packet Delivery Fraction (PDF)*: This is the total number of delivered data packets divided by the total number of data packets transmitted by all nodes. This performance metric gives us an idea of how well the protocol is performing in terms of packet delivery by using different traffic models.
- *Average End-to-End delay (Avg-End-to-End)*: This is the average time delay for data packets from the source node to the destination node. This metric is calculated by subtracting "time at which first packet was transmitted by source" from "time at which first data packet arrived to destination". This includes all possible delays caused by buffering during route discovery latency, queuing at the interface queue, retransmission delays at the MAC layer, propagation and transfer times.

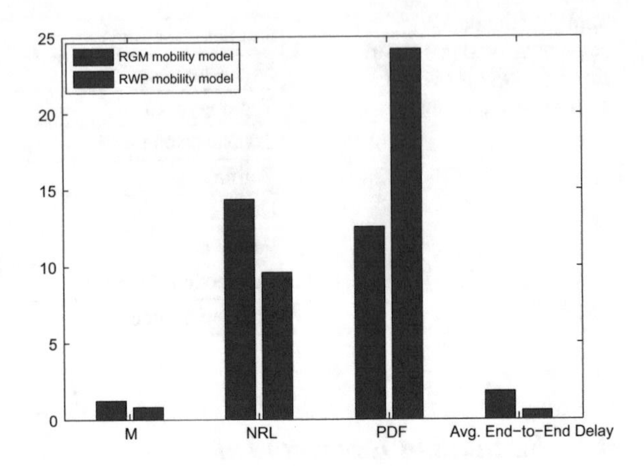

Fig. 1 OLSR protocol performance versus mobility metric M^{λ}

4.3 Results and Discussion

The Fig. 1 shows that with the same MANET parameters, the RWP mobility model witnesses less mobility compared to the RGM mobility model.

We find out that the three performance metrics are very good in the case of the RWP model, which guarantees less mobility: fewer messages of control ($NRL(RGM) > NRL(RWP)$), better delivery of data packets ($PDF(RGM) < PDF(RWP)$), and better average end to end delay *(Avg-End-to-End(RGM)>Avg-End-to-End(RWP))*. Moreover, we observe that the performance of the OLSR protocol is sensitive to the mobility in the MANET. Indeed, according to Fig. 1, we notice a slight difference between the two mobility measures, but for the three performance metrics, a very significant difference is found:

- *NRL(RGM)* $\simeq$ *3/2 NRL(RWP)*
- *PDF(RGM)* $\simeq$ *1/2 PDF(RWP)*
- *Avg-End-to-End(RGM)* $\simeq$ *3 Avg-End-to-End(RWP)*

Hence, so as to have the same performances of a protocol it is necessary to make simulations in the same *mobility condition*, i.e to chose scenarios with the same mobility measure before beginning simulations.

5 Conclusion

This paper presents a new metric of mobility measure for evaluating mobile ad hoc networks. This mobility metric is unified for all mobility models, and it is logical and lightweight. Moreover, we have study his behavior in different mobile ad hoc network configurations. Finally, we have applying this mobility metric to evaluate the performance comparison of the OLSR protocol through two different mobility models: RWP (based random) and RGM (with temporal dependence) models.

References

1. S. Cörson, J. Macker, Mobile Ad hoc Networking (MANET): Routing Protocol Performance Issues and Evaluation Considerations, Request or Comments 2501, IETF, (RFC 2501) (1999)
2. J. Jubin, J.D. Tornow, The DARPA packet radio network protocols. Proc. IEEE **75**(1), 21–32 (1987)
3. C.E. Perkins, E.M. Royer, S.R. Das, M.K. Marina, Performance comparaison of two on-demend routing protocols for ad hoc networks. IEEE Pers. Commun. **8**(1), 16–28 (2001)
4. B.-J. Kwak, N.-O. Song, L.E. Miller, A standard measure of mobility for evaluating mobile Ad Hoc network performance. IEICE Trans. Commun. E86-B(11), 3236–3243 (2003)
5. F. Bai, A. Helmy, A survey of mobility models in wireless Ad Hoc networks, in *Wireless Ad Hoc and Sensor Networks*, pp. 1–29 (2004)
6. T. Camp, J. Boleng, V. Davies, A survey of mobility models for Ad Hoc network research. Trends Appl. **2**(5), 483–502 (2002)
7. B. Divecha, A. Abraham, C. Grosan, S. Sanyal, Impact of node mobility on manet routing protocols models. J. Digit. Inf. Manag. (2007)
8. N. Pal, R. Dhir, Analyze the impact of mobility on performance of routing protocols in MANET Using OPNET modeller. Int. J. Adv. Res. Comput. Sci. Softw. Eng. **3**(6), 00 (2013)
9. X. Hong, M. Gerla, G. Pei, C.-C. Chiang, A group mobility model for Ad Hoc wireless networks, in *Proceedings of the 2nd ACM International Workshop on Modeling, Analysis and Simulation of Wireless and Mobile Systems, MSWiM 99* (ACM, USA, 1999), p. 5360
10. Y. Chenchen, L. Xiaohong, Z. Dafang, An obstacle avoidance mobility model, in *2010 IEEE International Conference on Intelligent Computing and Intelligent Systems (ICIS)*, vol. 3 (2010), pp. 130–134
11. N. Aschenbruck, E. Gerhards-Padilla, P. Martini, A survey on mobility models for performance analysis in tactical mobile networks (2008)
12. J. Broch, D.A. Maltz, D.B. Johnson, Y.-C. Hu, J. Jetcheva, A performance comparison of multi-hop wireless ad hoc network routing protocols, in *Proceedings of the Fourth Annual ACM/IEEE International Conference on Mobile Computing and Networking (Mobicom98)* (1998)
13. P. Johansson, T. Larsson, N. Hedman, B. Mielczarek, M. Degermark, Scenario-based performance analysis of routing protocols for mobile Ad-Hoc networks. pp. 195–206 (1999)
14. N. Sadagopan, F. Bai, B. Krishnamachari, A. Helmy, Paths: Analysis of path duration statistics and their impact on reactive manet routing protocols, in *MobiHoc03* (2003)
15. F. Bai, N. Sadagopan, A. Helmy,. Important: A framework to systematically analyze the impact of mobility on performance of routing protocols for ad hoc networks, in *Proceedings of IEEE Infocom* (2003)
16. X. Hong, T. Kwon, M. Gerla, D. Gu, G. Pei, A mobility framework for ad hoc wireless networks, in *Proceedings of ACM Second International Conference on Mobile Data Management (MDM 2001)* (2001)
17. B. Zhou, K. Xu, M. Gerla, Group and swarm mobility models for ad hoc network scenarios using virtual tracks, in *Proceedings of MILCOM*, pp. 289–294 (2004)
18. J. Boleng, W. Navidi, T. Camp, Metrics to enable adaptive protocols for mobile Ad Hoc networks, pp. 293–298 (2002)
19. T. Clausen, P.J. (eds.), C. Adjih, A. Laouiti, P. Minet, P. Muhlethaler, A. Qayyum, L. Viennot, *Optimized link state routing protocol (OLSR)Rfc 3626* (2003)
20. NS2, official Website. http://www.isi.edu/nsnam/ns/. Accessed 2017
21. F.J. ROS, *Um-OLSR version 8.8.0, University of Murcia, Spain.* http://masimum.dif.um.es/?software:um-olsr. Accessed 2017

Security Requirements and Model for Mobile Agent Authentication

Sanae Hanaoui, Jalal Laassiri and Yousra Berguig

Abstract Mobile Software Agent a technology that satisfies the requirements of intelligence in distributed systems; and has become increasingly used in serval domains but it is difficult to grantee its security because of its strong mobility over the network. And for today's Internet, the Secure Socket Layer (SSL) and Transport Layer Security (TLS) are the most widely deployed security protocols that used to exchange data over the Internet securely based on encrypted communication. The present paper seeks a solution to meet the security requirements of communication in Multi-Agent System (MAS) based on the JADE framework and by using cryptographical techniques to ensure a secure encrypted communication using SSL/TLS protocol (Paulson LC, Inductive Analysis of the Internet Protocol TLS, [1, 2]).

Keywords Authentication · SSL/TLS protocol · Mobile software agent · Security

1 Introduction

In all systems the guaranty of communication safety is the most important, which relies on the authentication process which is a process for identifying and verifying the identity of a subject [3]. In case the authentication stage failed the authenticated part wouldn't perform any operation for which it was given permission. The idea of an agent has entered the field of computer science and gained some acceptance but the security of this technology still a challenge and it has not obtained enough attention from the agent community [4]. Nevertheless, with the aim of agent technology to provide unrestricted secure solutions, security issues have to be addressed. In this

S. Hanaoui (✉) · J. Laassiri · Y. Berguig
Faculty of Science Kenitra Morocco, Systems and Optimization Laboratory, Ibn Tofail University, Kenitra, Morocco
e-mail: sanae.hanaoui@uit.ac.ma

J. Laassiri
e-mail: LAASSIRI@uit.ac.ma

Y. Berguig
e-mail: yousra.berguig@gmail.com

M. Elhoseny and A. K. Singh (eds.), *Smart Network Inspired Paradigm and Approaches in IoT Applications*, https://doi.org/10.1007/978-981-13-8614-5_11

paper we are only concerned with the security of the system, especially securing the agent authentication based on the protocol ssl tls using cryptographical techniques.

The agents may engage in variant MAS platform providers. In the objective of protecting both the agent owner and the receiver platform, it must be guaranteed their integrity and adequate trust level during the migration process [5, 6]. And if we take on consideration the security requirement for mobile agent systems, authentication is a very important stage. Before the secure transfer of a mobile agent between the two platforms, they must authenticate each other for mobile agent systems authentication refers to a process in which the platform ensures that the other platform in the communication is in fact who it is declared to be [7].

The remainder of this paper is organized as follows. Section 2 briefly investigates the security problematic in the mobile agent systems and exposes various communication threats on mobile agents also explores some security requirements to protect it. In Sect. 3 we will discuss the protocol ssl tls and some of its characteristics and our motivation for using this protocol. Section 4 discusses and elaborates the background of research. While Sect. 5 gives detailed information about the authentication approach adopted to secure our agent. In Sect. 6 we elaborate the implementation of the approach. Finally, Sect. 7 conclude the paper.

2 Mobile Agents Security Overview

the mobile agent faces critical, security risks because of its strong mobility, where its code, data and state are exposed to other platforms in which it migrates for getting information or execution of a goal which gives to either malicious platform or other agent a chance to alter or even kill the agent before it attains its goal or before it accomplishes its assigned task. Therefore the following security properties should be taken on consideration [8–10] so that the agent will be more certain about the visited platform and vice-versa for the visited platform:

- Authentication and Authorization: by assuring that communication initiate from its originator.
- Privacy and Confidentiality: by assuring confidentiality of exchanges and interactions in a mobile agent system in order to secure the communication of a mobile agent with its environment.
- Non-repudiation: by logging important communication exchanges to prevent later denials
- Accountability: by recording not only unique identification and authentication but also an audit log of security-relevant events, which means all security-related activities must be recorded for auditing and tracing purposes. In addition, audit logs must be protected from unauthorized access and modifications.
- Availability: Agent platform should be capable of detecting also recovering from software and hardware failures. It should be able to deal with Dos attacks as and to prevent it as well.

- Anonymity: the mother platform should keep agent's identity hidden from other agents and maintains anonymity so when necessary and legal to the determine agent's identity.
- Fairness or trust: the necessity to ensure fair agent platform interaction where the agent should be able to assess the trustworthiness of information received from another agent or from an agent platform.

3 SSL/TLS Discussion

to create a secure link between a server and a client machine over the internet, the likely advised secure communications protocol is the Secure Sockets Layer (SSL) and Transport Layer Security (TLS) protocols [11, 12].

3.1 Secure Sockets Layer Protocol

It's a security protocol on the Internet that is used to encrypt connections between two parties [13], most commonly between a web browser and a web server. It's mostly referred to SSL by the dual moniker SSL/TLS, since the protocol suite was upgraded and renamed to Transport Layer Security back in 1999. The goal of SSL was to provide secure communication using classical TCP sockets with very few changes in API usage of sockets to be able to leverage security on existing TCP socket code.

The SSL/TSL protocol empowers applications to be only as secure as the underlying infrastructural components. SSL/TLS is a detached protocol that inserts itself between the application protocol (generally HTTP, but any other is perfectly possible) and the transport protocol (TCP). By acting as such, TLS demand very few changes to the protocols below and above, so the protocol can operate nearly transparently for users, indicating that users need not to be aware of the fact that the protocol is in place. Of course this comes with some drawbacks [14].

SSL/TLS has evolved considerably since its beginnings. Nowadays SSL 2.0 and 3.0 are considered insecure, so they are being replaced by the newer TLS 1.0/1.1/1.2 versions. The latest standard version is TLSv1.2 http://tools.ietf.org/html/rfc5246, while the upcoming TLS v1.3 is still in the draft stage.

3.2 Description of the Protocol

The TLS protocol placed between the Application Layer and the Transport Layer. It is divided into two main sublayers. This is the general structure of the protocol (see Fig. 1):

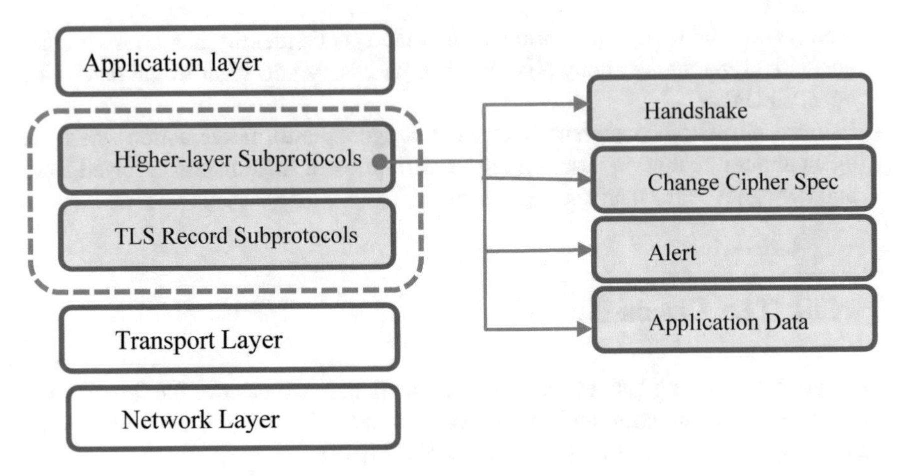

Fig. 1 The general structure of the protocol SSL/TLS

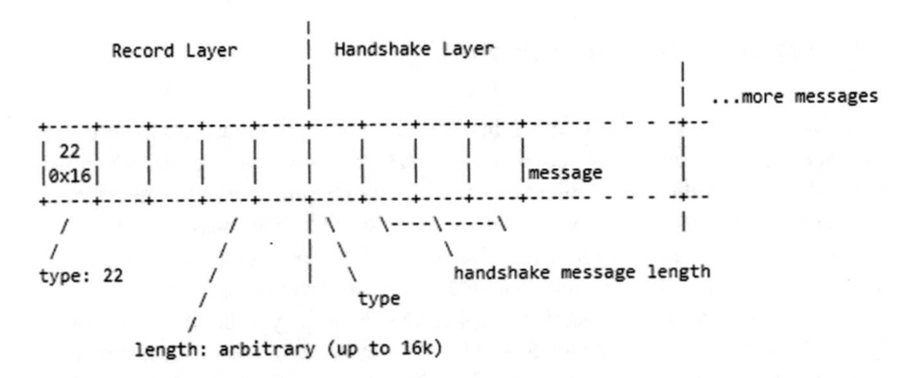

Fig. 2 Handshake protocol format

The higher layer is stacked on top of the SSL Record Protocol, and comprises four subprotocols. Each of these protocols has a very precise purpose, and are used at different stages of the communication [13]:

- Handshake Protocol: It allows the peers to authenticate each other and to negotiate a cipher suite and other parameters of the connection. The SSL handshake protocol involves four sets of messages that are exchanged between the client and server. Each set is typically transmitted in a separate TCP segment (see Fig. 2).
- ChangeCipherSpec Protocol: It makes the previously negotiated parameters effective, therefore the communication becomes encrypted.
- Alert Protocol: Used for communicating exceptions and indicate potential problems that may compromise security.
- Application Data Protocol: It takes random data (application-layer data generally), and send it through the secure channel.

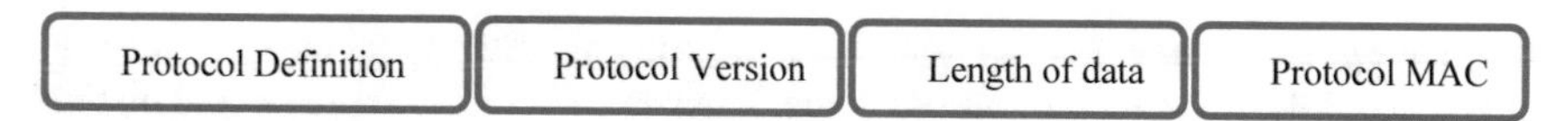

Fig. 3 The header of record protocol

The lower layer is stacked on top of TCP, as it is a connection-oriented and reliable transport layer protocol. This layer made of basically of the TLS Record Protocol.

The TLS Record header comprises three fields, necessary to allow the higher layer to be built upon it (see Fig. 3).

3.3 Motivation

We opted for SSL/TLS because it provides authentication (Signature Authentication) also confidentiality and integrity. However, TLS provides a more secure method for managing authentication and exchanging messages [15, 16]. And because the authentication of the agent is less developed by the jade community. However, for the JADE-S platform requires that users (the owner of the agent or the container) must be authenticated, by providing a username and password, in order to be able to perform instructions on the platform. But it doesn't authenticate a mobile agent itself so that the visited platform verifies the identity of the arrived agent to ensure that the agent has not become malicious as a consequence of alterations to its state.

4 Related Works

The research in the area of mobile agents' programming is still active specially in the security of this technology. Several projects have developed execution environments for mobile agents. However, the authentication mechanisms have been partially addressed:

Vila et al. [17] they have explored the challenges, issues, and solutions to fulfil the security requirements of a Multi-Agent System (MAS) based on the JADE framework. By presenting a sufficient security vision for MAS. Thus, several security features are considered, from the authentication over encryption of the exchanged data up to the authorization of the access to services assigned only to a determined group. For Bayer and Reich [5] they have addressed specific security requirements for mobile software agents one of the requirement is the authentication and possible threats for agent system operations in the context of Java Agent Development. Their main objective was to show existing vulnerabilities and security breaches by analyzing the security of JADE platform, giving existing improvements of the confidentiality of software agents merging from one agent platform to another and introducing trusted agents and their implementation in JADE. While Berkovits et al. [18]

they have granted three security goals for mobile agent systems and have proposed an abstract architecture to achieve those goals. Their architecture is based on four distinct trust relationships between the principals of mobile agent systems. They have used existing theory—the distributed authentication theory of Lampson et al.—to clarify the architecture and to show that it meets its objectives. Ismail, Emirates [19] they have described authentication mechanisms for mobile agents. In these mechanisms, the authentication of mobile agents is controlled by the mobile-agents platform using digital signature and a Public Key Infrastructure. Agents are authenticated via the authentication of their running platforms.

5 Proposed Approach

The authentication approach that we propose, is based on Signature Authentication where signature authentication or public key authentication is an alternative method of identifying yourself to a login server, instead of typing the password. In our case is to identify our agent to the visited platform or server based on the protocol ssl/tls. therefore we propose that both the platform and agent identify each other at the arrival to the destination platform by requesting authentication using the protocol tls (see Fig. 4).

5.1 *Description of the Approach*

The scenario is as follows. When a mobile agent calls the migration method by then the agent server or the hosted platform as we proposed in our first paper [20]

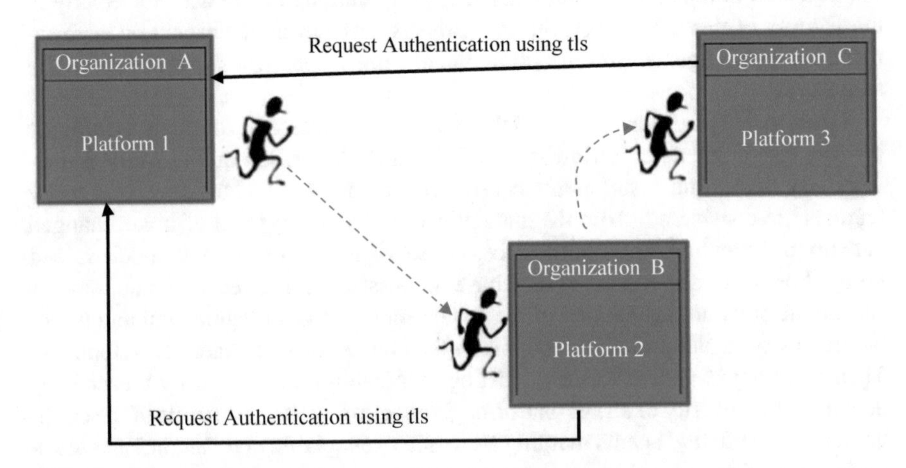

Fig. 4 Authentication process

Fig. 5 The agent header

encrypts a formatted header using RSA algorithm (see Fig. 5) that consists of the owner id, code for the agent's permissions and agent id also a signature of data or code and it signs the agent code and state digitally in an easily seen way to the agent using a private key. The agent (state and code) along with its header is sent to the destination agent server or to the destination platform which is identified by its node IP address and its port number. On reception of the signed agent and its header, the receiver agent server (the visited host) request an authentication using the protocol tls to verifies the identity of the mobile agent platform which sends the agent. The receiver agent server then decides whether to accept or reject the reception of the agent based on the successful authentication and digital certification verification of the agent platform (see Fig. 6).

6 Experimentation: Implementation on Jade

In this section, we present an architectural overview of a mobile agent scenario followed by a description of our implementation of the authentication approach. This implementation is conducted in JADE and we are limiting our simulation on containers located in the same physical platform.

The practical tests of the implementation are carried out in a Machine which contains two containers that will represent the destination machines. The technical characteristics of the machine are: Intel Core i3 processor has 2.3 GHz with 4 GB of RAM. For the creation, management, mobility and execution of Agents we adopt JADE Snapshot from version During the Trip of the agent from the native platform to the hosting.

6.1 Architectural Overview

Each agent platform or server in the system communicates with a trusted third-party certification authority (CA) to obtain a private key and its corresponding digital certificate. The agent server's digital certificate is digitally signed by the CA. A KeyStore is associated with each agent platform or server in the system. which is used to store and manage private keys along with their corresponding digital certificates. The hosted platform or agent server retrieves its private key and the corresponding digital certificate from the KeyStore to sign a mobile agent and its header. The signed

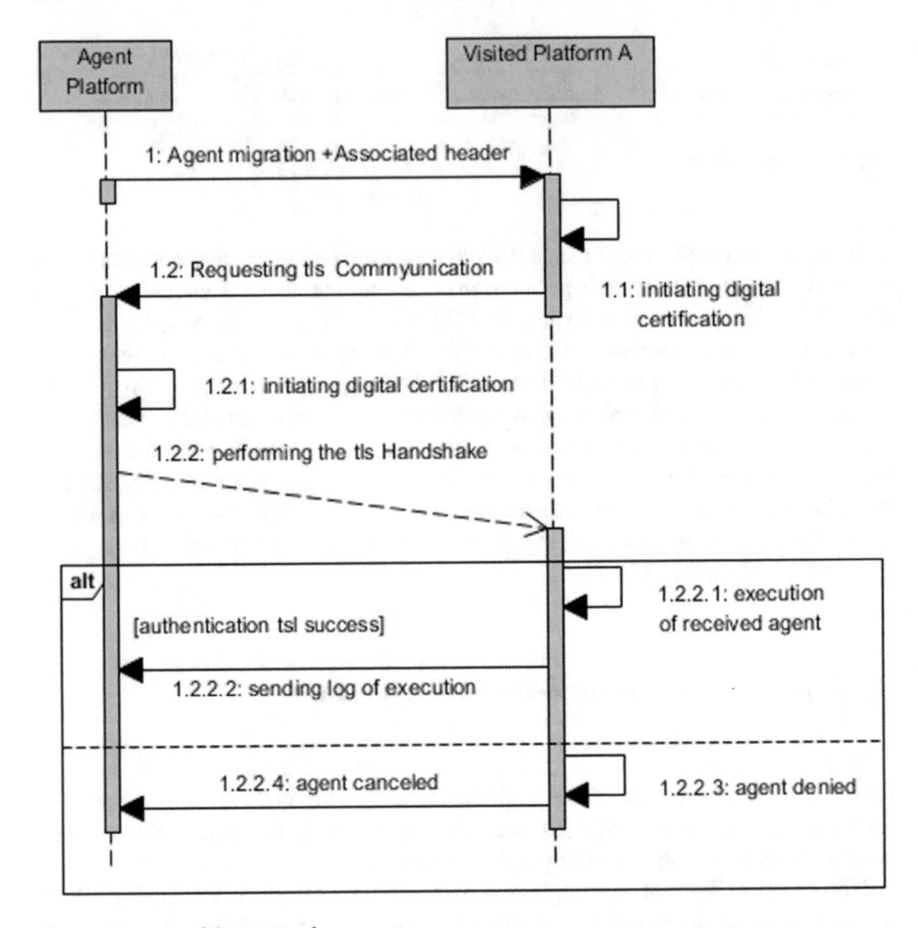

Fig. 6 Scenario of the approach

mobile agent associated with its header is then sent to the destination platform. On reception of the mobile agent associated with its header, the visited platform initiates a tls connection to retrieves the CA's digital signature and to decrypt the mobile agent's header and to verify the agent signature in order to make sure that the right agent does the right things.

To summarize the design of our minimal implementation of a secure mobile agent authentication, an agent migration to a destination agent server consists of the following steps (see Fig. 7):

a. Serialization the state of the assigned agent.
b. Serialization of the agent state and associating the agent with the header.
c. Retrieval of the agent platform public key and digital certificate from the local KeyStore.
d. Creation of an object signature to be used in signing the agent.
e. Initialization of the signature object with the server private key.

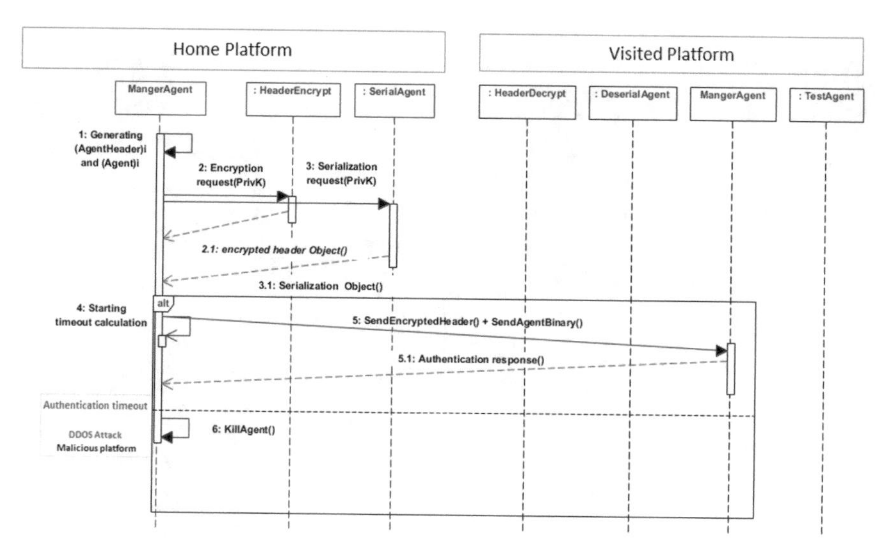

Fig. 7 Authentication scenario

f. Updating the signature object using the agent's state for encoding.
g. generation of the mobile agent signature using the signature object from step d).
h. generation of our agent header then encrypting our header using the private key which includes the signature and sender agent
i. associating the header to agent's state, code, and sending of the agent header and the agent to the destination agent server
j. Reception of the agent in the destination agent Server or platform and creation of a new thread for the execution of the agent.
k. Initiating the communication using the tls protocol to Retrieval of the sender agent server's public key and digital certificate
l. De-serialization of the agent's state.
m. Decrypt the header using the CA's public key shared by the protocol tls after verification of the sender agent server's digital certificate using the CA's public key from step k)
n. Verification of the agent's signature.
o. Run the agent if verification succeeds.

6.2 Simulation

We will adopt a concrete illustration of the adopted authentication scenario by consider a mobile agent that visits the Web sites of several prepharmacies searching for a plan that meets a customer's requirements. We focus on four servers: a customer server, a prepharmacies server, and two servers owned by competing prepharmacies,

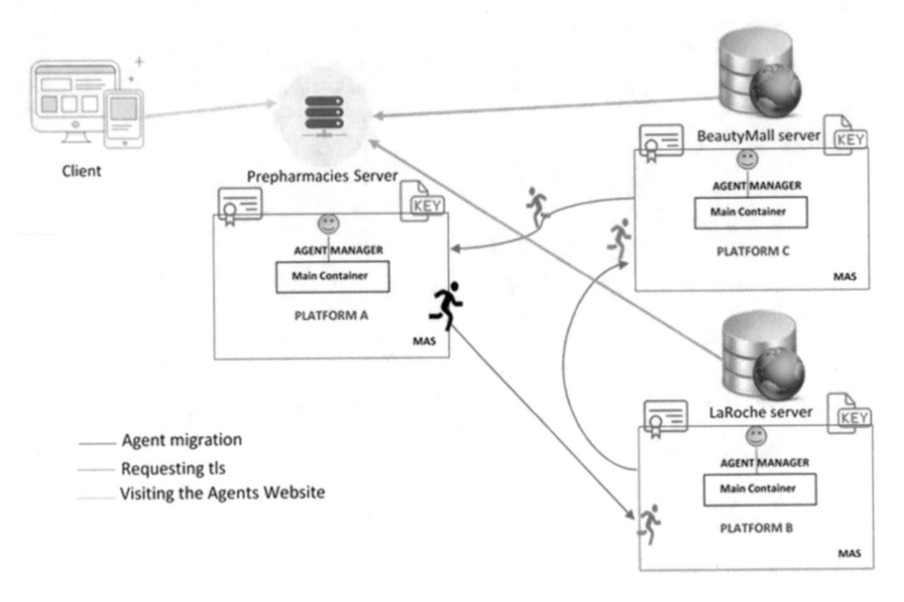

Fig. 8 Simulation architecture in JADE platform

for instance beautyMall and LaRoche. The mobile agent is programmed by a web site for selecting the best prepharmacy price. The agent Manager dispatches the agent to the beautyMall server where the agent queries the product database. With the results stored in its environment, the agent then migrates to the LaRoche server where again it queries the product database. The agent compares product and price information, decides on a plan, migrates to the appropriate prepharmacy server, and reserves the desired product. Finally, the agent returns to the customer with the results (see Fig. 8).

7 Conclusion

In this paper, we explored some security requirement for mobile agent also we presented the most used secure communication protocol ssl/tls and we gave our motivation for using this protocol in our authentication approach. Our Approach is based on digital signatures and public key infrastructure shared with the communication protocol tls. In this model, we assumed that an agent would trust other agents if they would come from a trusted agent server since an agent server which executes an agent has a full control over that agent. In addition, our authentication approach is adequate for a mobile agent platform or on a server, since an agent has to be signed every time it is sent. In this way the verification of the non-repudiation also the integrity of the agent is met, accordingly it is easier to identify a malicious agent platform that sends a malicious agent. The agent platform from which we receive an agent is responsible to verify the integrity of the agent from a previous sender. Our authentication approach ensures portability as available mobile agents can use the secure platform with no changes in the agents' application codes.

References

1. L.C. Paulson, *Inductive Analysis of the Internet Protocol TLS*
2. W. Goralski, Securing sockets with SSL. in *The Illustrated Network, Morgan Kaufmann*, pp. 585–606 (2009)
3. N. Constantinescu, C.I. Popirlan, *Authentication Model Based on Multi-agent System*, vol. 38, no. 2, pp. 59–68 (2011)
4. G. Stoneburner, *Underlying Technical Models for Information Technology Security* (Gaithersburg, MD, 2001)
5. T. Bayer, C. Reich, *Security of Mobile Agents in Distributed Java Agent Development Framework (JADE) Platforms Security of Mobile Agents in Distributed Java Agent Development Framework (JADE) Platforms,* no. April, 2017
6. S. Bijani, D. Robertson, A review of attacks and security approaches in open multiagent systems. Artif. Intell. Rev. **42**(4), 607–636 (2014)
7. S. Alami-Kamouri, G. Orhanou, S. Elhajji, Overview of mobile agents and security. Proc. 2016 Int. Conf. Eng. MIS, ICEMIS (2016)
8. M. Kaur, S. Saxena, *A Review of Security Techniques for Mobile Agents*, pp. 807–812 (2017)
9. N. Borselius, *Security in Multi-Agent Systems*, no. April 2014
10. B. Amro, *Mobile Agent Systems, Recent Security Threats and Counter Measures*
11. J. Liang, J. Jiang, H. Duan, K. Li, T. Wan, J. Wu, When HTTPS meets CDN: a case of authentication in delegated service, in *Proceedings - IEEE Symposium on Security and Privacy* (2014) pp. 67–82
12. K. Bhargavan, C. Fournet, M. Kohlweiss, A. Pironti, P. Y. Strub, Implementing TLS with verified cryptographic security, in *Proceedings - IEEE Symposium on Security and Privacy* (2013) pp. 445–459
13. P. Kocher, Internet Engineering Task Force (IETF) A. Freier Request for Comments: 6101 P. Karlton Category: Historic Netscape Communications (2011)
14. A. Castro-Castilla, Traffic analysis of an SSL/TLS session—The Blog of Fourthbit (2014). http://blog.fourthbit.com/2014/12/23/traffic-analysis-of-an-ssl-slash-tls-session. Accessed 11 Aug 2018
15. IETF, "full-text," *Internet Eng. Task Force* (2018)
16. S. Turner, Transport layer security. IEEE Internet Comput. **18**(6), 60–63 (2014)
17. X. Vila, A. Schuster, A. Riera, Security for a multi-agent system based on JADE. Comput. Secur. **26**(5), 391–400 (2007)
18. S. Berkovits, J.D. Guttman, V. Swarup, Authentication for mobile agents. Mob. Agents Secur. LNCS **1419**, 114–136 (1998)
19. L. Ismail, U.A. Emirates, Authentication Mechanisms for Mobile Agents Leila Ismail United Arab Emirates University (2007)
20. S. Hanaoui, Y. Berguig, J. Laassiri, *On the security Communication and Migration in Mobile Agent Systems* (Springer, Cham, 2019) pp. 302–313

Internet of Thing for Smart Home System Using Web Services and Android Application

Khaoula Karimi and Salahddine Krit

Abstract A smart home is a home based on the internet of thing to enable the control and the remote monitoring of home's devices and to allow the user to adapt the system to his desires and needs. This paper presents an approach to implement a smart home system using the Internet of thing IoT, Web services, and an Android App. The proposed model focuses on (1) An Arduino Uno Wi-Fi platform for interoperability among sensors, actuators, and communication protocols. (2) The REST framework makes the home appliances accessible and connected also it improves data exchange. (3) An Android App providing several functionalities by which the user can control the home device from anywhere. We present the smart home architecture and its application in a use case. Our goal is providing a low cost, effective and low-cost smart home system which can be controlled easily from anywhere.

Keywords Smart home · Smartphone · Internet of Things (IoTs) · Web services · Android App

1 Introduction

With the speed of lifestyle evolution, technological development, and the High-speed internet access, researches aim to create a solution which connects all objects to the internet and provide to the users a simple platform to control this objects. This solution is called the Internet of Thing.

Internet of Things (IoT) consist of devices that allow all objects to be connected from anywhere and anytime [1]. It's a set of elements that connect and share vast amounts of security data [2]. Currently, there are 3.4 billion Internet users. Reports indicate that by 2020, there will be 50 billion connected devices worldwide [3].

K. Karimi · S. Krit (✉)
Laboratory of Engineering Sciences and Energies, Polydisciplinary Faculty of Ouarzazate, Ibn Zohr University, Ouarzazate, Morocco
e-mail: salahddine.krit@gmail.com

K. Karimi
e-mail: karimi.khaoula92@gmail.com

M. Elhoseny and A. K. Singh (eds.), *Smart Network Inspired Paradigm and Approaches in IoT Applications*, https://doi.org/10.1007/978-981-13-8614-5_12

The closest concept to the field of Internet of Thing is the smart home. It can be defined as a house in which all appliances can interact and be connected to the internet.

The Smart Home system uses the Internet of Thing technologies to provide homeowners with comfort, security, and interaction with its appliances from anywhere [4]. The connection between these appliances is done through a microcontroller getaway. For the proposed smart home system we have used an Arduino board. It works for any IoT applications design and, can update programs according to the needs [5].

To allow the user to control devices remotely the present work gives an Android application that will be installed on a smartphone. Several interfaces have been developed to ensure the control and oversight of the house. The exchange data between the Android app and the web server we have to use web services. In the proposed architecture RESTful has based web services that make the user communicate the actuators and sensors directly and in real time. It's more easy and lightweight to talk between machines. And simple HTTP is used to make calls between devices [6]. RESTful applications use HTTP requests to post data (create and/or update), read data (e.g., make queries), and delete data. Thus, REST uses HTTP for all four CRUD (Create/Read/Update/Delete) operations [7].

Rest of the paper is organized as follows: Sect. 2 describes the proposed architecture for our smart home and its technologies. Section 3 we provide a scenario of implementation of Smart home system using REST full and Arduino Uno and Sect. 4 concludes this paper.

2 Related Work

The Internet of Thing definition is still vague and unspecified due to its coverage of several domains [8] and its employ and use in many constructs of our living style [9].

For the authors [10] IoT is a system that can collect information from many environments even if their generation and through several devices. Also it can be defined as a sit of smart devices that are connected to internet and exchange data [11].

Home automation is the most practical area for the implementation of an IoT platform. Since the smart home system is intelligent sensors and devices interconnected in order to ensure comfort and safety to the inhabitant [12].

Due to the smartphone evolution and the access to the internet, the remote control of the smart home devices is based on using a smartphone application. Several models of smart homes have been implemented.

The [13] proposed a system that allows user to take decision and control light, set alarms, and reminders using an android application. This system is based on a Raspberry Pi 3 processor, an API of a messaging website to send SMS to the owner and a camera that take photos when the doorbell is rung and send it to user via SMTP.

The authors [14] implement Smart Home System based on Raspberry Pi using Wi-Fi, Internet of Things, and an Android Application. He aims to create a user interface for desktops or laptops and applications for mobiles and tablets. But the proposed system had some drawbacks like the Raspberry Pi requires at least 2 h updating itself for starting controlling.

Bhatnagar et al. [15] developed a system using Node MCU ESP 8266 12E as the controller, Firebase database and Wi-Fi as the communication protocol. Also, it could be controlled using an android application or Google Assistant.

Recently the concept of web services is being used for bringing the devices to the web. The most used technologies for Web services are REST and SOAP.

Muthulakshmi and Latha [16] defined the SOAP Simple Object Access Protocol as an exchange protocol for a decentralized environment. The author provides the reason for choice of SOAP, is that SOAP allows messages exchange using applications and interoperable services and without the knowledge of underlying systems.

On the other side, the authors [17] aim to provide a solution for enabling the interoperability on a higher level in building automation systems using RESTful-based web services. The principle is to use the same well-known techniques used in traditional web applications to avoid some drawback.

3 The Proposed Architecture for Smart Home/Smartphone System

3.1 System Architecture

The smart home's system must allow users to measure home conditions monitor and remotely control home appliances. The proposed system is based on 4 major elements: Arduino Uno Wi-Fi as a microcontroller, sensors for measuring home condition, actuators for monitoring directly home appliance, and an Android app for controlling the smart home using a smartphone. Figure 1 illustrates a general overview of the proposed architecture.

In order to make the system architecture of our smart home more understandable, Fig. 2 shows the components of the system and the interaction between them.

- Data store: Stores and analysis data (commands or data reading) from Android App and sensors.
- Sensors: measure home conditions and send this data to the microcontroller.
- Actuators: receives commands transferred by the Arduino Uno Wi-Fi for realizing certain actions.
- Android App: an App which can be downloaded and installed on a smartphone. It allows user to send commands and controls the system remotely.
- Web Service: it registers the commands from the Android app in the database, checks the current state of the actuators, and it returns a response on the commands performed by the user.

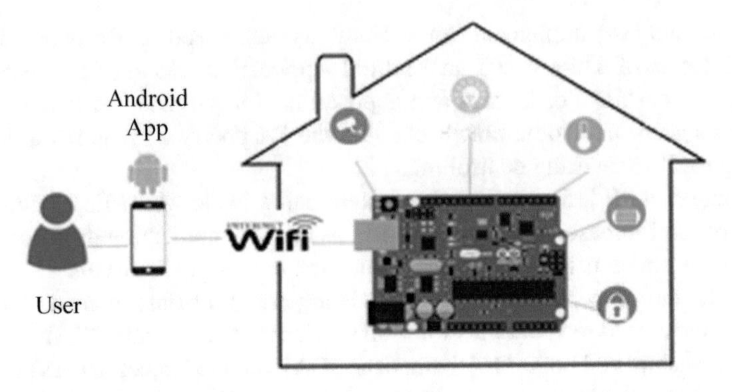

Fig. 1 An overview of the proposed architecture

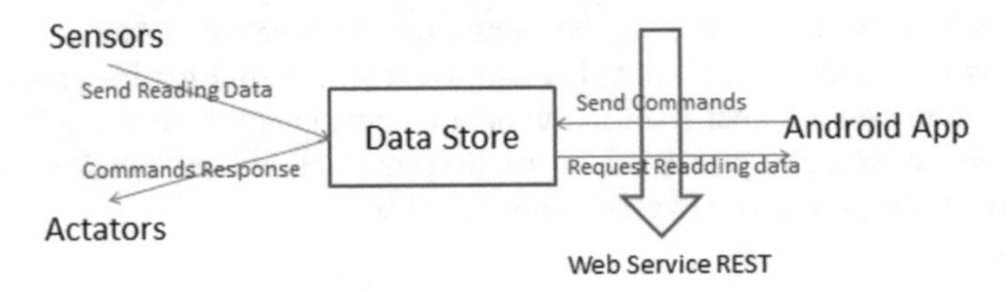

Fig. 2 Components of our home automation

3.2 Technologies for the Smart Home Implementation

To realize our smart home system we used several software and hardware technologies. This section describes and discusses more about them.

3.2.1 Arduino Uno Wi-Fi

The Arduino Uno Wi-Fi is used to implement the microcontroller of our smart home system. It's an Arduino board with an integrated Wi-Fi module. It is easy to program and it is implemented to provide communication between devices and technologies. The Arduino Uno Wi-Fi is based on the ATmega328P with an ESP8266 Wi-Fi Module integrated. It has 14 digital input/output pins (of which 6 can be used as PWM outputs), 6 analogue inputs, a 16 MHz [18].

3.2.2 Web Service REST

Using Web Services is the most interoperable and easy way to provide access to remote services or to allow applications to communicate with each other. Representational State Transfer (REST) and Simple Object Access Protocol (SOAP) are two kinds of web services. But Rest-based web services are simple compared to

Device	@	Action

Fig. 3 The general form of command sent from the Android app

SOAP and it offers an effective way of interacting with lightweight clients, like smartphones. The proposed system has RESTful-based Web-services that transmits requests by HTTP for many resources using the operations POST and GET, and it often returns a response. The web server responses for REST can be delivered in multiple formats. JSON (JavaScript Object Notation) is often used, but XML is also valid. In the proposed system we have used JSON because it is independent of other languages, and it is easy to parse and generate codes. The JSON syntax is expressed in “name”:“value”.

At the Android application level, getting JSON code from the URL sent from the server is ensured by the following function:

```
//import…..
public static String getJsonFromServer(String url) throws
IOException
  {
      BufferedReader inputDate = null;
      URL jsonUrl = new URL(url);
      URLConnection dc =  jsonUrl.openConnection();
      dc.setConnectTimeout(6000);
      dc.setReadTimeout(6000);
      inputData = new BufferedReader(new  InputStreamReader(
                 dc.getInputStream()));

      //read JSON results into a string
      String jsonResult = inputStream.readLine();
      return jsonResult;
}
```

When the user monitors the home devices from the proposed app, the command is sent to the microcontroller server through internet. The general form of these commands Fig. 3 is chosen in such a way that the server can easily read the command and execute it to the actuators. For example, if the user wants to turn ON the air-conditioner, the command form sent will be airconditioner@ON.

Figure 4 shows some examples of interaction between The Android app and the server web via Rest using JSON format for response messages.

3.2.3 Android App

An Android application was developed using the ADT (Android Developer Tools). Java programming language using the Android Software Development Kit has been used for the implementation and the development of the smart home app. Android Studio, which is the official Integrated Development Environment (IDE) for Android

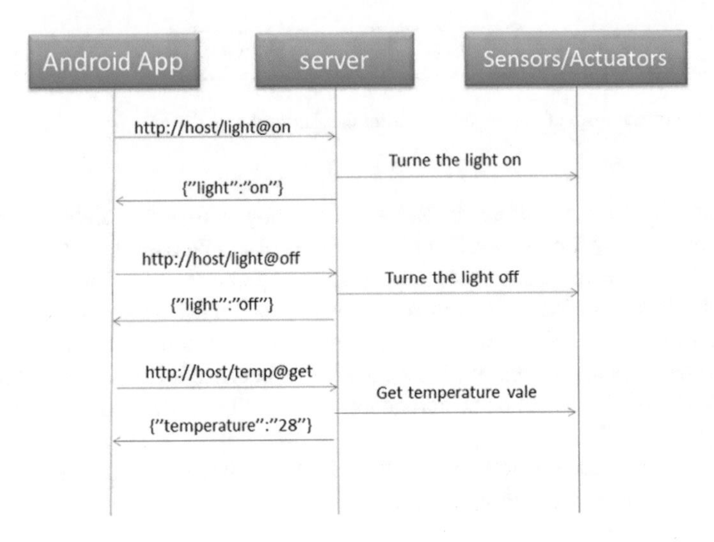

Fig. 4 Communication between Android app and Web server using REST

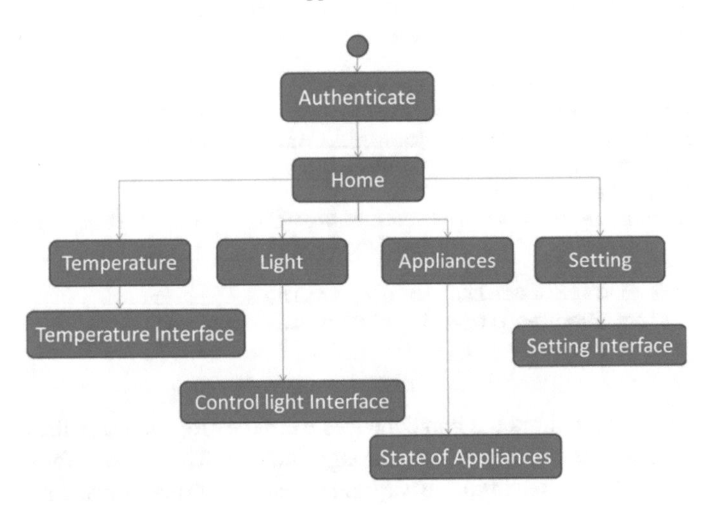

Fig. 5 Functionalities of the proposed Android app

app development, has been used to develop the smart home app. The main activities of the smart home app are presented in Fig. 5.

First, the user should enter his login and password to authenticate. If these elements are incorrect, a message invites him to check the login and the password. When the authentication was correct the app shows the main interface as shown in Fig. 6a which allows the user to:

- Controlling the home's temperature.
- Controlling the lights in all rooms.

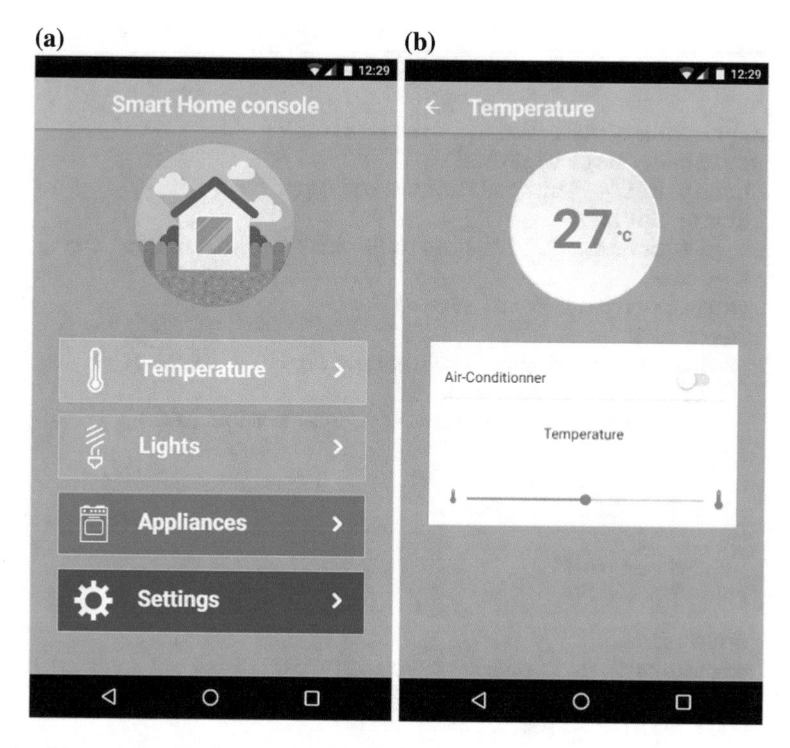

Fig. 6 **a** The main interface, **b** temperature controlling interface

- Getting the state of all connected devices such as Camera, humidity sensor….
- Setting or getting more information about the system.

For example, when the user clicks on controlling temperature, this interface as shown in Fig. 6b provides the user the ability to:

- Getting the temperature of the room.
- Turn ON/OFF the Air-conditioner
- Changing the level of temperature

The temperature value is displayed on a button. It is updated by clicking on this button.

To get this value on the interface we have parsing the JSON data coming from the micro web server using the following class for android code source:

```
//import …
public class TemperatureActivity extends AppCompatActivity
{
    AsyncTask<Void, Void, Void> DoTask;
    String JSONStringFORM;
    String url = "http://host/temp@get ";
    Button button;
button = (Button) findViewById(R.id.TemperaturBoutton);
DoTask = new
   AsyncTask<Void, Void, Void>() {
     @Override
     protected Void doInBackground(Void...params) {
        try {
        JSONStringFORM = getJsonDataFromUrl(url);
            }
        catch (IOException e) {
        e.printStackTrace();
            }
        return null;
     }

     @Override
     protected void onPostExecute(Void result){
       super.onPostExecute(result);
       //Parsing JSON Data
       try {
       JSONObject jsonobj = new JSONObject(JSONStringFORM);
       //JSONStringFORM = {"temperature":"37"}
       button.setText(jsonobj.getString("temperature") +
" °C");     }
       catch(Exception e){
       e.printStackTrace();
           }
     }
        };
DoTask.execute();
}
```

We have used AsyncTask which is called a UI Thread, to allow us a background treatment on our Android app without slowing down the navigation, and update the interface of the application at the end of treatment.

In the background the class calls the getJesonDataFromUrl function to retrieve data JSON as string and in Postecxute we have created a function that parses the Data JSON and displays it on the interface.

4 Scenarios of Implementation of Smart Home System Using RESTfull and Arduino UNO WI-FI

We implemented different scenarios using the web services and IoT. In this section, we are going to present the implementation of an example of these functionalities which is controlling light by using motion detector in the room.

The light control is provided by using the motion detector PIR to detect the presence of someone at the room then it sends commands to the microcontroller. Arduino board get the information and decide to turn NO the light if there is someone at the room and also if the time is between 8 pm and 7 am. Figure 7 illustrates the electric block diagram and the implementation of this scenario.

An excerpt from the code source that is written on the Arduino IDE to realize this scenario is as follows:

```
void loop(){
  value = digitalRead(PIRpin);
  if (value == HIGH)
    {
    digitalWrite(lEDPin, HIGH);
    delay(150);
       if (pirState == LOW)
       {
       Serial.println("mention true");
       pirState = HIGH;
       delay(5000) ;
       }
    }
```

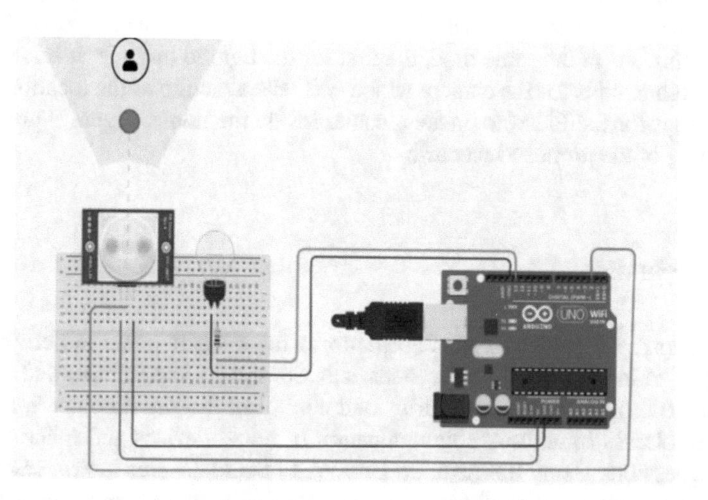

Fig. 7 Electric block diagram and the implementation of the light control using a motion sensor

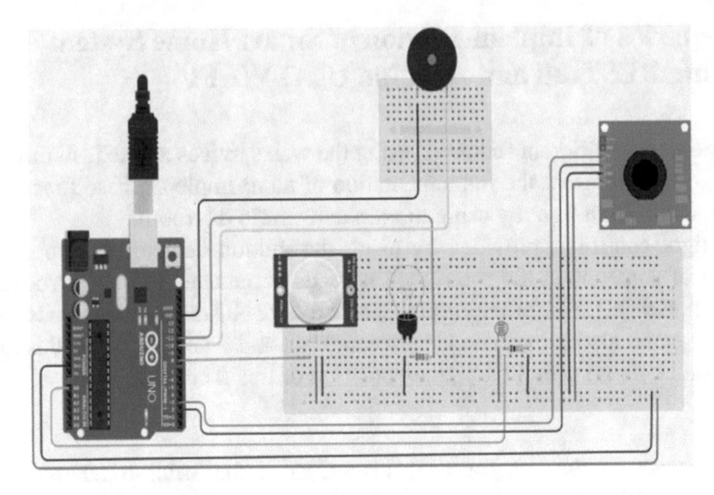

Fig. 8 Electric block diagram for gardens security

```
    else
      {
      digitalWrite(lEDPin, LOW);
         if (pirState == HIGH){
       Serial.println("mention false ");
       pirState = LOW;
       delay(500) ;
     }
 }
```

This scenario can be used in all rooms of the house. Also, it can be used and with the addition of a camera and a piezoelectric speaker, as is shown in Fig. 8 for more security in the gardens. In fact, the PIR will detect the presence of a person then send three commands at the same time, the first for the light to turn ON if it is night, the second command is for the camera which will take a picture of the location, and the third is sent for the Piezo to create a sound inside the home. Figure 9 presents the processing of the proposed scenario.

5 Conclusion

In this paper, we present a flexible platform for a smart home system based on Internet of Thing IoT, and REST web services. Several devices (sensors and actuators) are used to provide comfort, security, and control of the smart home. We used the RESTful classes to address communication of home devices and make the smart home accessible. Using the Arduino Uno Wi-Fi board ensures interoperability and communication among appliances. In the proposed model we provide an Android app for controlling and monitoring the smart home from anywhere and time. Any

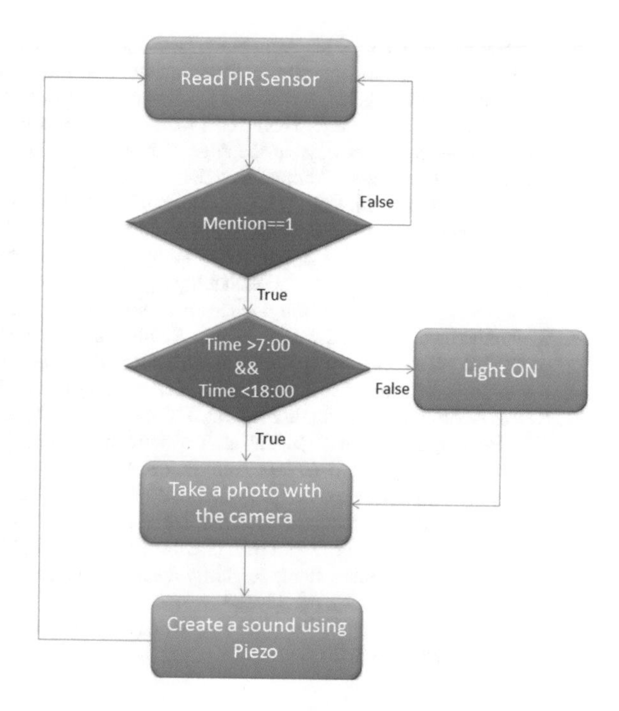

Fig. 9 Web based motion detector and camera control security

Android smartphone is able to download the app and be used to remotely control the smart home devices. Future works will focus on providing a more secure system that detects any threat or intruder and informs the owner of this house in real time.

References

1. S. Feng, P. Setoodeh, S. Haykin, Smart home: cognitive interactive people-centric Internet of Things. IEEE Commun. Mag. **55**(2), 34–39 (2017)
2. A. Dorri, S.S. Kanhere, R. Jurdak, P. Gauravaram, Blockchain for IoT security and privacy: the case study of a smart home, in *2017 IEEE International Conference on Pervasive Computing and Communications Workshops (PerCom Workshops)* (IEEE, 2017), pp. 618–623
3. S. Bhadoria, R. Oliva Ramos, Raspberry Pi 3 home automation projects (2017)
4. T. Chakraborty, S.K. Datta, Home automation using edge computing and Internet of Things, in *2017 IEEE International Symposium on Consumer Electronics (ISCE)* (IEEE, 2017), pp. 47–49
5. S. Wadhwani, U. Singh, P. Singh, S. Dwivedi, Smart home automation and security system using Arduino and IOT. Int. Res. J. Eng. Technol. **5**(02), 1357–1359 (2018)
6. S. Bhadoria, R. Oliva Ramos, Raspberry Pi 3 home automation projects (2018)
7. S. Kim, J.Y. Hong, S. Kim, S.H. Kim, J.H. Kim, J. Chun, Restful design and implementation of smart appliances for smart home, in *2014 IEEE 11th International Conference on Ubiquitous Intelligence & Computing and 2014 IEEE 11th International Conference on Autonomic & Trusted Computing and 2014 IEEE 14th International Conference on Scalable Computing and Communications and Its Associated Workshops* (UIC-ATC-ScalCom) (IEEE, 2014), pp. 717–722

8. R. Basatneh, B. Najafi, D.G. Armstrong, Health sensors, smart home devices, and the Internet of Medical Things: an opportunity for dramatic improvement in care for the lower extremity complications of diabetes. J. Diabetes Sci. Technol. **12**(3), 577–586 (2018)
9. A.M. Rahmani, T.N. Gia, B. Negash, A. Anzanpour, I. Azimi, M. Jiang, P. Liljeberg, Exploiting smart e-Health gateways at the edge of healthcare Internet-of-Things: a fog computing approach. Future Gener. Comput. Syst. **78**, 641–658 (2018)
10. X. Zheng, Z. Cai, Y. Li, Data linkage in smart Internet of Things systems: a consideration from a privacy perspective. IEEE Commun. Mag. **56**(9), 55–61 (2018)
11. R. Söderström, M. Westman, Internet of Things: The smart home (2018)
12. K. Karimi, S. Krit, Systems and technologies for smart homes/smart phones, in *Proceedings of the Fourth International Conference on Engineering & MIS 2018—ICEMIS '18* (2018)
13. K. Pampattiwar, M. Lakhani, R. Marar, R. Menon, Home Automation using Raspberry Pi controlled via an Android application. Int. J. Curr. Eng. Technol. (2017)
14. P. Rathod, S. Khizaruddin, R. Kotian, S. Lal, Raspberry Pi based home automation using Wi-Fi, IOT & Android for live monitoring. Int. J. Comput. Sci. Trends Technol. (IJCST) **5**(2) (2017)
15. H.V. Bhatnagar, P. Kumar, S. Rawat, T. Choudhury, Implementation model of Wi-Fi based smart home system, in *2018 International Conference on Advances in Computing and Communication Engineering (ICACCE)* (IEEE, 2018), pp. 23–28
16. A. Muthulakshmi, R. Latha, The SOAP based mechanism for home environment using web services. Electr. Comput. Eng. Int. J. (ECIJ) **3**(2), 53–60 (2014)
17. H. Järvinen, P. Vuorimaa, Interoperability for web services based smart home control systems, in *WEBIST*, no. 1 (2014), pp. 93–103
18. Store.arduino.cc, Arduino Uno Wi-Fi [Online] (2018), https://store.arduino.cc/arduino-uno-wifi. Accessed 28 Sept 2018

Bat with Teaching and Learning Based Optimization Algorithm for Node Localization in Mobile Wireless Sensor Networks

G. Kadiravan and Pothula Sujatha

Abstract Mobile wireless sensor networks (MWSNs) have become popular as it replaces the conventional wireless sensor networks to networks with movable sensor nodes which have the capability to observe different environmental scenarios. And, the mobile sensor nodes have the nature of changing their location often in the provided sensing region. The process of node localization in MWSN is a challenging task which intends to determine the location coordinates to every device with unknown positions in the target region. Presently, different metaheuristic based optimization algorithms for node localization have been devised. This paper introduces a BAT-TLBO algorithm by altering the characteristics of traditional bat algorithm (BA) with the teaching and learning based optimization (TLBO) algorithm for determining the proper node localization in the network. Furthermore, the BAT-TLBO algorithm has been tested with different scenarios with varying anchor node density. The simulation results reported that, on average, the localization error of BA, MBA and BAT-TLBO are 0.259, 0.541 and 0.219 respectively. From the experimental results, it is ensured that the BAT-TLBO algorithm showed better localization performance than the compared algorithms in terms of robustness, improved localization success ratio and computation time.

Keywords Node localization · Metaheuristic · Echolocation · Anchor node

1 Introduction

Mobile wireless sensor networks (MWSNs) act as a major part in practical scenarios in that the sensor nodes are movable and not stagnant in nature. MWSNs are flexible to a great extent when compared to WSNs since the sensor nodes undergone deployment in different environments and it manages with fast topology modifications [1]. Mobile sensor nodes comprise of a microcontroller, variety of sensors to capture the environmental conditions (moisture, acoustics, temperature, movement, etc.), a

G. Kadiravan (✉) · P. Sujatha
Department of Computer Science, Pondicherry University, Puducherry, India
e-mail: kathirkadir@gmail.com

M. Elhoseny and A. K. Singh (eds.), *Smart Network Inspired Paradigm and Approaches in IoT Applications*, https://doi.org/10.1007/978-981-13-8614-5_13

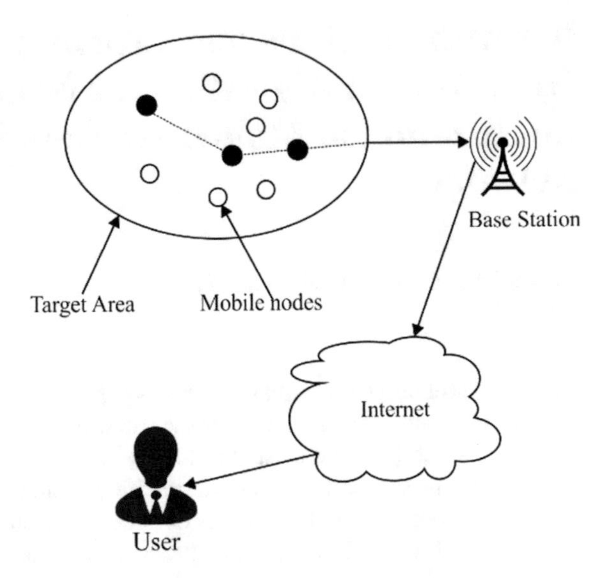

Fig. 1 Architecture of MWSN

radio transceiver, in addition it is energized through an inbuilt power unit [2]. The significant applications of MWSNs are ecological surveillance, weather forecasting, seismic surveillance, acoustic recognition, hospitals, navy, military, smart buildings, and data gathering platforms [3]. There are two pairs of disputes to MWSNs; hardware and software. The major hardware restrictions are inbuilt battery source, compact size, and installation cost. In addition, the nodes in MWSN should achieve high energy efficiency for the longer network lifetime [4]. At the same time, the complexity level of the algorithms derived for various tasks like localization, clustering, routing and so on should be low [5].

The WMSN comprises of a number of mobile sensor nodes which sense the environment and transmits the sensed data to the base station (BS). The basic architecture of a MWSN is shown in Fig. 1. Actually, the nodes in MWSN are modeled with many supplementary sensors to sense the environment, microcontroller, storage area, communication unit, analog to digital converter (ADC) and power supply. Yet again, the nodes are restricted with inbuilt memory, battery source, processing and radio facility because of its tiny size [6]. And, the mobile sensor node structure is nearly analogous to the actual sensor node. In addition, few supplementary units are included in mobile sensor nodes like location finder like GPS, mobile unit, and power generator. The structure of the node in MWSN is depicted in Fig. 2. Location finder unit is employed to recognize the coordinates of the sensor node and the mobile unit makes the sensor node movable. The power generator unit is liable to produce energy for satisfying additional energy needs of the sensor node using explicit methods like solar cell.

In a few more task like that, node localization acts a key enabling role. Node localization is needed to report the initialization of events, helps to request information from specific node, data transmission and so on. Hence, node localization in

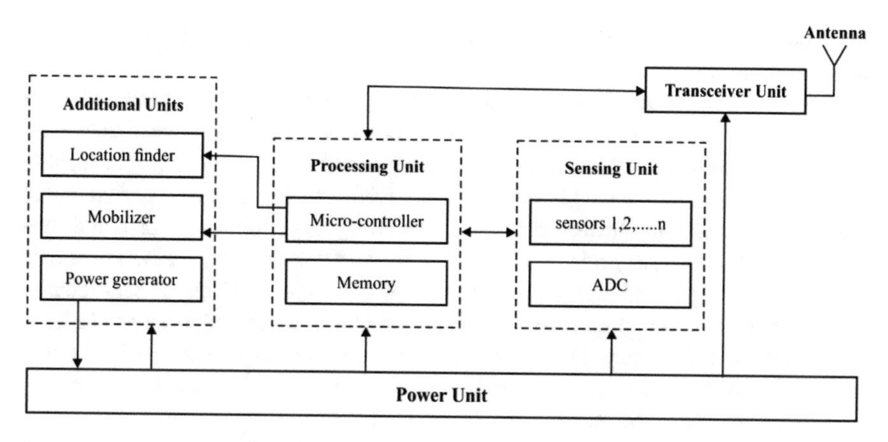

Fig. 2 Architecture of a node in MWSN

MWSN is considered as a crucial design issue [7]. In a network, there would be a huge count of sensor nodes closely dispensed at locations that might be movable. In most of the MWSN, the data grouped through these sensors would be pointless if not the position out off wherever the data is acquired is well-known. This creates the localization capability of a network more important [8, 9]. Hypothetically, a localization determining gadget like GPS might be able to be employed for a sensor to position itself. Although, it is unrealistic to employ GPS in each sensor node as a MWSN comprises of more number of nodes and GPS would be highly expensive. Notwithstanding, GPS does not work ever in indoor environment, so alternate solutions must be used [10]. To resolve the issue, numerous localization approaches have been evolved. In spite of availing GPS to every node in MWSN, the localization methods simply consider that the minimum number of nodes contain GPS hardware. These nodes are frequently known as anchor nodes and they can easily recognize its locations [11]. In other way, some localization algorithms can be employed to determine the distance between the anchors and the actual sensor able to converse with some close by sensors and it examine distances between them by employing few localization algorithms (for example, received signal strength (RSS), time of arrival (ToA)) and then obtain its location using the distance [12–16]. In a standard point of view, the process of optimization can be described to find the best solution of the function from the system within constraints [17–20].

MWSN is considered as a multi-model and multidimensional optimization issue and it can be resolved by population dependent stochastic methods. Some localization algorithms using Genetic Algorithm (GA) are employed in [21, 22], which the estimated optimum node positions of every one-hop neighbor nodes. A two phase centralized localization method which employs Simulated Annealing (SA) Algorithm and GA is introduced in [23]. Particle Swarm Optimization (PSO) based approach is projected in [12, 24], to improve the localization accuracy. Each of these heuristic and meta-heuristic optimization algorithms is dominant for resolving the localization issue.

Most of the localization algorithms are developed based on the natural activity of biological systems and/or physical systems. For instance, PSO is proposed depending upon the swarm activities of birds and fish in addition SA depends on annealing process of metals whereas GA was stimulated from through natural activities like inheritance, mutation, selection, and crossover. Every algorithm has its pros and cons. A bat algorithm (BA) is projected in [25] depend on echolocation activities of bats. BA has been introduced stimulated through the fantastic activities of echolocation of micro bats. BA to a great extent efficient compared to other approaches in terms of precision and efficiency. The BA suffers from low success rate due to the fact that bat is not capable to discover each and everyway in the search space. Consequently, to conquer this issue, the existing BA is changed.

In this paper, we propose a BAT-TLBO algorithm by combining the nature of BA and teaching and learning based optimization (TLBO) algorithm [26]. In the algorithm, the movement of bats is changed with the characteristics of the TLBO algorithm, to discover the optimal solution in which direction wherever the narrow movement of bat cannot find the solution. The BAT-TLBO technique is superior when compared to the BA and modified BA (MBA) by means of computational speed and mean localization error (MLE) due to the nature of BAT-TLBO algorithm which improves the exploration of search space in an effective way. The BAT-TLBO algorithm has been tested with different scenarios with different anchor node density and the results are compared with BA and MBA for node localization in MWSN.

The succeeding part of the paper arranged as follows: Sect. 2 describes the proposed BAT-TLBO algorithm and the application of BAT-TLBO algorithm on MWSN Localization is discussed in Sect. 3. Section 4 provides the simulation outcome and discussion. Finally, the Sect. 5 concluded the paper with future enhancements.

2 Bat Algorithm

Bats are interesting animals and its inherent characteristic of echolocation has drawn interest from more number of researches in various domains [27]. The principle of echolocation follows the nature of sonar: bats, particularly micro-bats, produce a loud and short pulse of sound which strikes to an object and then an echo will be returned back to their ears [28]. By this way, the bats determine the actual distance of the object [29]. Additionally, this remarkable orientation method helps the bats to differentiate between obstacle and prey, allows it for hunting the prey without any light [30]. Inspired from the nature of the bats, Yang [31] has proposed a metaheuristic optimization algorithm named as BA. This algorithm has been proposed to act as a group of bats for identifying the prey/foods by the use of echolocation property. To formulate the bat algorithm, the following rules should be followed [31]:

- Every bat utilizes the echolocation principle for sensing distance and they can distinguish between the food/prey and obstacles in a supernatural manner;
- A bat b_i flies in a random manner with a velocity v_i at position x_i with a predefined frequency $f_{\min}$, differing wavelength (λ) and loudness (A_0) for searching the prey. It can routinely modify the λ of the emitted pulses and vary the rate of pulse emission $r \in [0, 1]$, based on the nearness of their target;
- Though the A_0 can fluctuate in different forms, Yang [31] assumed that the A_0 lies in the range of large (positive) A_0 to a minimum constant value A_{min}.

At the beginning, initialization takes place for every bat in terms of starting position x_k, velocity v_k and frequency f_k are initialized for every bat b_k. For every time step t, T is the maximum number of iterations, the motion of bats is calculated by the updation of v_k and position by the use of Eqs. (1)–(3).

$$f_k = f_{min} + (f_{max} - f_{min})\beta \tag{1}$$

$$v_k^t = v_k^{t-1} + \left(x_k^t - x^*\right) f_k \tag{2}$$

$$x_k^t = x_k^{t-1} + v_k^t \tag{3}$$

where indicates a randomly generated number lies in the range of [0, 1], x_k^t indicates the value of the decision variable l for a bat l at the time step t. The result of f_k can be employed to is used to manage the pace and range of the movement of the bats. The variable x^* indicates the present global best location (solution) for decision variable l, which is attained when compared to all the solutions given by m bats.

3 Proposed Node Localization Algorithm in MWSN

3.1 BAT-TLBO Algorithm

The traditional bat algorithm is altered by the TLBO algorithm. The pseudo code of the BAT-TLBO algorithm is illustrated in Algorithm 1. The new solutions are created by the equations inspired from the nature of knowledge exchange among teacher and students in the learning duration. This algorithm initiates with the group of agents (known as classes) where every agent (known as student) is a significance solution to the problem. The TLBO has two stages: teaching stage and learning stage. In the former stage, the best agent of the class is chosen as the teacher and remaining agents (students) tries to enhance the knowledge level through learning from teacher. In the latter stage, the student tries to develop the knowledge level through interaction with each other. This nature is used in the node localization problem, the present current node namely δ_k arbitrarily selects other node δ_l, where $k \neq l$, and when the selected node has better objective value, the present node moves toward it, or else, the δ_k moves away from δ_l, which can be defined by,

$$\delta_i^{update\_location} = \begin{cases} \delta_k + r.(\delta_k - \delta_l) iff(\delta_k) < f(\delta_l) \\ \delta_k + r.(\delta_l - \delta_k) iff(\delta_k) \geq f(\delta_l) \end{cases} \quad (4)$$

where r is a random number lies in the range of 0 and 1. When δ_k gives a better objective value, it is substituted with the present agent, else δ_k will remain.

Algorithm 1: Pseudo code for BAT-TLBO algorithm	
1.	Objective function $f(x), x = (x_1, \ldots\ldots\ldots x_d)^T$
2.	Initialize the bat population $x_k(i = 1,2,3 \ldots\ldots n)$ and v_k
3.	Define Pulse frequency f_k at x_k
4.	Initialize pulse rates r_k and the loudness A_k
5.	While (t< Max number of iterations)
6.	Generate new solutions by adjusting frequency and updating velocities and locations/solutions according to equation (1), (2) and (3).
7.	Evaluate objective function. If solution is away from the optimal value of objective function then move to step 8 else jump to step 9.
8.	Generate new solutions by following TLBOalgorithm given in Eq.(4)
9.	If $(rand > r_k)$
10.	Select a solution among the best solutions
11.	Generate a local solution around the selected best solution
12.	End if
13.	Generate a new solution by flying randomly
14.	If $(rand < A_k \& f(x_k < fx^*))$
15.	Accept the new solutions
16.	Increase r_k and reduce A_k
17.	End if
18.	Rank the bats and find the current best x^*
19.	End while
20.	Post process results and visualization

3.2 *Proposed Node Localization Algorithm in MWSN*

A single hop range based distributed approaches can be employed to localize the nodes in MWSN localization, to determine the coordinates of many sensor nodes by the use of anchor nodes [32]. For determining the location of *N* sensor nodes, the following process will take place. At the beginning, *N* sensor nodes undergo random deployment in the sensing filed in the C-shaped topology. In a network of *N* node, *M* anchor nodes are present, which know their coordinates earlier, undergo deployment in C-shape topology. $(N - M)$ are the unknown nodes, whose coordinates needs to be determined. Every node has a transmission range of R. An unknown node can easily determine its own location coordination if and only if it has more than two non-coplanar anchor nodes as neighbors, i.e. the node is known to be a localizable node. Every localizable node computes its distance from every one of the nearby anchor nodes. The distance measurement gets distorted with Gaussian noise n_i because of environmental considerations.

$$\hat{d}_L = [d_k + n_k] \quad (5)$$

where d_i is the distance from the localizable node to anchor node which is determined by

$$d_k = \sqrt{(x - x_k)^2 + (y - y_l)^2} \quad (6)$$

where (x, y) is the location coordinates of the unknown node and (x_k, y_k) is the location coordinates of the ith anchor node in the nearby area. The location determination of a given unknown node can be formulated as an optimization problem, involving the minimization of an objective function indicating the localization accuracy. So, every unknown node needs to determine the location coordinates execute the stochastic algorithms in an independent manner for the localization by identifying the coordinates (x, y). The objective function for localization problem is defined as:

$$f(x, y, z) = \frac{1}{M}\sum_{k=1}^{M}(d_k - \hat{d}_L) \quad (7)$$

where $M \geq 3$, (2D position of a node requires at least three anchor nodes) in communication range, R, of the unknown node. The process of localization is a repetitive task. The unknown nodes with a minimum of 3 nearby anchor nodes are localized and then it is also considered as the anchor nodes to help the localization process of the remaining unknown nodes. This procedure is iterated till all the unknown nodes are correctly localized. Every algorithm provides the optimal coordinates of the unknown nodes, i.e. (x, y) by minimizing the error function. The localization error can be termed as the distance between the actual and calculated coordinates of an unknown node which is determined as the mean of square root of distance of computed node coordinates (x_i, y_i) and the real node coordinates (X_i, Y_i), for $i = 1, 2, \ldots N_L$ (where N_L is the number of localized nodes) as equated in Eq. (8).

$$E = \frac{\sum_{i=M+1}^{N}\sqrt{(x_i - X_i)^2 + (y_i - Y_i)^2}}{(N_L)} \quad (8)$$

Localization error is defined in terms of mean localization error (MLE) in a transmission range R to verify the performance of the application results, which can be computed as

$$E = \frac{\sum_{i=M+1}^{N} \sqrt{(x_i - X_i)^2 + (y_i - Y_i)^2}}{(N_L)R} \tag{9}$$

4 Performance Evaluation

For validating the performance of the proposed BAT-TLBO algorithm, it is simulated in MATLAB R2014a on a PC of 8 GB RAM. To simulate the BAT-TLBO algorithm, the unknown and anchor sensor nodes are arbitrarily place in a C-shape topology in the sensing region. The communication range of every sensor node is found to be identical. For comparison purposes, the BA with MBA algorithm is used. The BAT-TLBO has been tested with different scenarios with varying anchor node density.

4.1 Parameters Setup

To authenticate the highlights of the BAT-TLBO algorithm, a comparison is with state of art methods by executing the objective function for $P \times I$ times where P represents the population size and I indicate the maximum number of iterations (optimum or $P \times I$ time reached, whichever is earlier). Here, P and I are kept as 20 and 100 respectively for all the applied algorithms. For these algorithms, the extra control parameters f_{min} and f_{max} are kept as 0.01 and 0.05 kHz respectively. The starting values of the variables r and A are set to 0.5 ms and 0.2 ms, respectively. The parameter initialization of the MWSN scenario is tabulated in Table 1.

Table 1 Parameter initialization

Parameter	Value
Node count (n)	200
Anchor nodes (m)	30
Sensing field	200 * 200 m^2
Ranging error	5%
Transmission range	30 m
Population size (P)	20
Number of iterations (I)	100
f_{min} and f_{max}	0.01 and 0.05 kHz
Pulse rate (r)	0.5 ms
Loudness (A)	0.2 ms

4.2 Results and Discussion

Figures 3, 4, 5 and 6 shows the node localization results of the BA, MBA and the proposed BAT-TLBO algorithm. Here, blue color indicates the localized sensor nodes, red color represents the anchor nodes and the black circle indicates the unknown nodes.

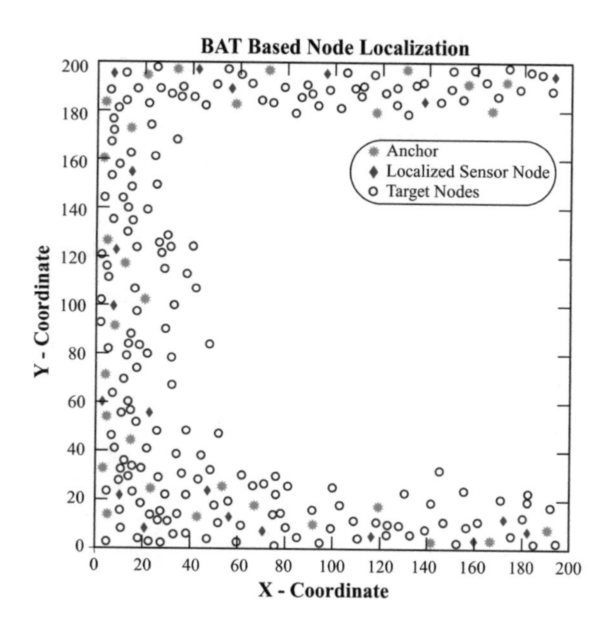

Fig. 3 Node localization using BA

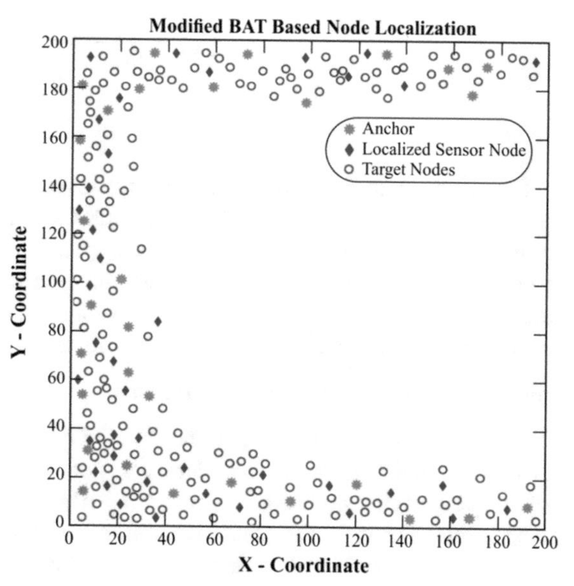

Fig. 4 Node localization using MBA

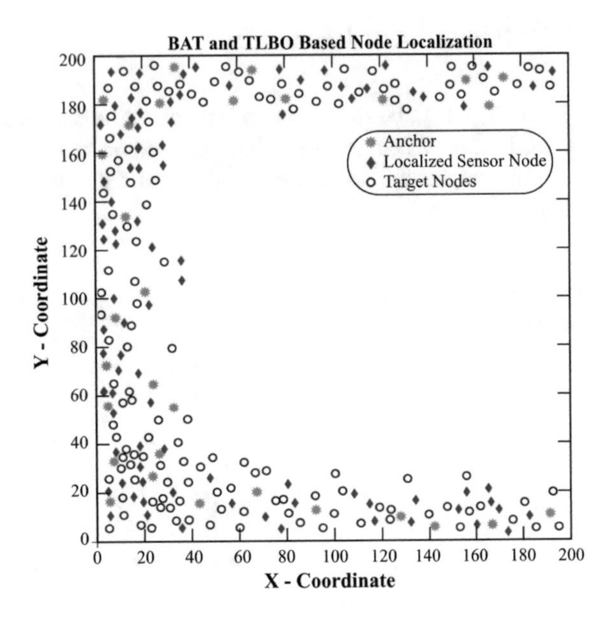

Fig. 5 Node localization using BAT-TLBO

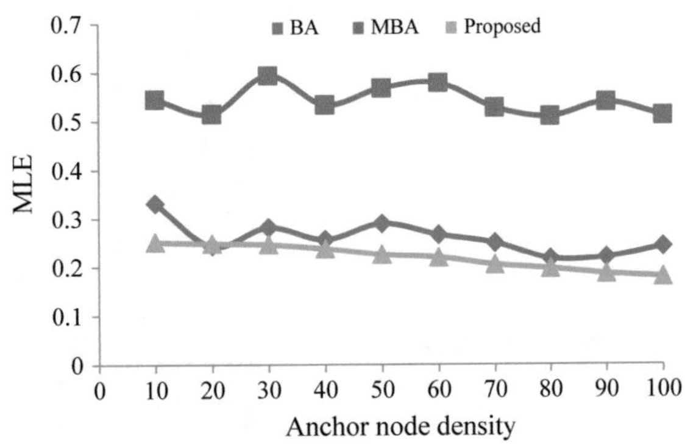

Fig. 6 Impact of varying anchor node density in terms of MLE

From these figures, it is observed that only less number of nodes is localized using BA compared to MBA and Bat-TLBO algorithms. At the same time, the number of localized nodes (NL) of MBA is high compared to BA. But, the maximum NL is obtained by the BAT-TLBO algorithm than the compared ones. Moreover, the proposed BAT-TLBO algorithm achieved the highest success rate due to the fact that the number of nodes which are localized is higher compared to BA and MBA.

Impact on Varying Anchor Node Density

The performance of the output parameters of MWSN node localization is determined for varying anchor node density. Here, 200 nodes are deployed in the sensing field with varying anchor node density from 10 to 100. The high anchor node density will be beneficial due to the fact that many references will be available for the unknown nodes. The number of nodes which can be localized is mainly based on the count of

Table 2 Impact of varying anchor node density

Anchor	BAT			M-BAT			Proposed		
	MLE	Time (s)	NL	MLE	Time (s)	NL	MLE	Time (s)	NL
10	0.3314	10.14	6	0.5445	0.88	172	0.251	0.71	180
20	0.2427	10.92	7	0.5144	0.90	177	0.248	0.73	185
30	0.2809	11.98	7	0.5932	0.86	179	0.246	0.76	187
40	0.2561	13.50	8	0.5326	0.98	180	0.237	0.78	188
50	0.2888	13.00	9	0.5669	0.82	182	0.225	0.81	192
60	0.2658	12.00	10	0.5774	0.98	188	0.219	0.82	194
70	0.249	13.76	12	0.5264	0.79	188	0.204	0.85	196
80	0.2163	12.55	13	0.5084	0.87	193	0.196	0.87	197
90	0.2191	12.98	14	0.5374	0.86	194	0.185	0.93	199
100	0.2416	13.29	20	0.5101	0.96	195	0.179	0.97	201

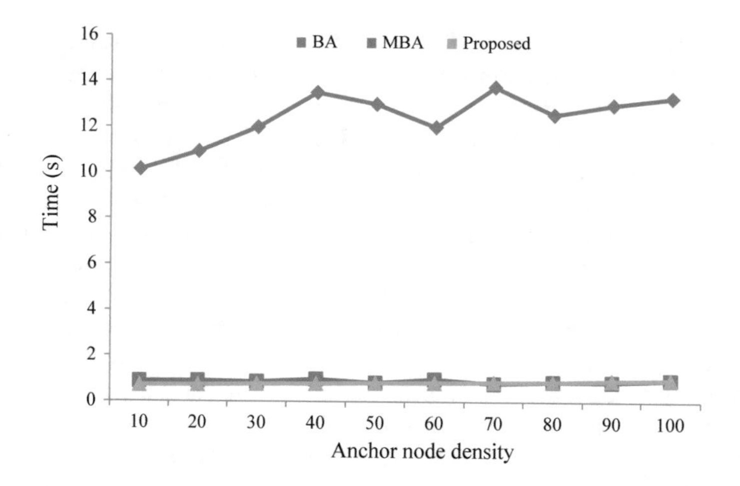

Fig. 7 Impact of varying anchor node density in terms of computation time

anchor nodes in the network. To study the impact of anchor nodes on MWSN node localization performance, an experiment is carried out using different anchor node density. The measures employed to validate the performance are MLE, computation time and NL. The comparison results of these measures are given in Table 2. And, Figs. 6, 7 and 8 depicts the comparative analysis of three localization algorithms namely BA, MBA, and BAT-TLBO algorithms in terms of MLE, computation time and NL respectively.

Figure 6 shows the results of different versions of bat algorithms in terms of MLE with respect to different number of anchor node density. From the graph, it is observed that the MLE starts to decrease with increasing anchor node density. From this figure, MBA achieved worst performance with a maximum MLE of 0.51. On average, the

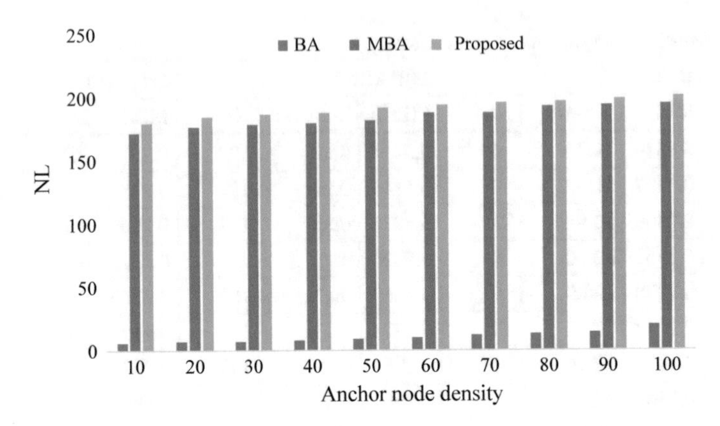

Fig. 8 Impact of varying anchor node density in terms of NL

MLE of BA, MBA and BAT-TLBO are 0.259, 0.541 and 0.219 respectively. The proposed BAT-TLBO algorithm obtained the superior performance with the lowest MLE of 0.179. At the same time, the BA showed better performance than the MBA with a minimum MLE of 0.2416. But, the BAT-TLBO outperforms the compared methods in a significant way. The incorporation of TLBO in the localization process of the BAT-TLBO leads to the higher level of precise localization performance.

Figure 7 provides an investigation of the impact of the varying anchor nodes in terms of computation time. From the figure, it is evident that the computation time is too high for the BA compared to MBA and proposed algorithm. However, the MBA consumes less time for the localization process than BA. At the same time, proposed BAT-TLBO consumes slightly lesser time than the MBA. It is also noted that the computation time starts to increase slightly with the increasing anchor node density. At the 100th anchor node, the BAT-TLBO requires a computation time of 0.97 s whereas the BA and MBA needs a computation time of 13.29 s and 0.96 s respectively.

The comparison results of the influence of the localization performance on the varying anchor node density in terms of the NL are shown in Fig. 8. As seen in figure, for 10 anchor nodes, the BAT-TLBO properly localizes 180 nodes whereas the BA localizes only a minimum of 6 nodes. Though MBA attains a NL of 172 which is higher than the results obtained by the BA, it fails to show superior performance over the BAT-TLBO algorithm. Similarly, for the 100 anchor nodes, the BAT-TLBO effectively localizes 201 which are much higher than the NL of the other algorithms. The order of increased number of NL of different algorithms is BAT-TLBO, MBA and BA. For better understanding, the Table 3 reported the MLE of the different algorithms on the 30 iterations.

Impact on Ranging Error

Table 4 tabulates the performance of the node localization algorithms on the different ranging error. In this simulation, the ranging error is set in the range of 5–50 with

Table 3 MLE for 30 iterations

Trails	BA	MBA	Proposed
1	0.2397	0.5364	0.2165
2	0.2922	0.548	0.2154
3	0.1948	0.5385	0.2132
4	0.167	0.5373	0.2097
5	0.2447	0.4753	0.2078
6	0.1935	0.6188	0.1912
7	0.1964	0.5515	0.1931
8	0.2414	0.5766	0.1912
9	0.1632	0.5391	0.1908
10	0.4429	0.5297	0.2467
11	0.1455	0.5476	0.1378
12	0.2282	0.575	0.2134
13	0.2315	0.5529	0.2098
14	0.1845	0.5727	0.1793
15	0.2244	0.537	0.2132
16	0.1736	0.527	0.1654
17	0.2844	0.5753	0.2486
18	0.1611	0.591	0.1543
19	0.1215	0.561	0.1098
20	0.1649	0.499	0.1542
21	0.3415	0.5685	0.2789
22	0.2516	0.526	0.2278
23	0.1845	0.5426	0.1643
24	0.2244	0.5503	0.2087
25	0.1736	0.5476	0.1487
26	0.2844	0.575	0.2367
27	0.2282	0.5297	0.2086
28	0.1935	0.5475	0.1865
29	0.2315	0.5753	0.2076
30	0.2397	0.5364	0.2165

an interval of 5%. The ranging error is a crucial parameter, the quantity of Gaussian noise linked with the distance measurements which greatly influences the precision of localization. The dependency of MLE on ranging error is analyzed for 30 iterations for every value of error ranges between 5 and 50%. Figures 9, 10 and 11 depicts the comparative analysis of three localization algorithms namely BA, MBA, and BAT-TLBO algorithms in terms of MLE, computation time and NL respectively on ranging error.

Table 4 Impact of ranging error

Ranging error (%)	BA			MBA			Proposed		
	MLE	Time (s)	NL	MLE	Time (s)	NL	MLE	Time (s)	NL
5	0.1260	5.22	23	0.5191	0.89	197	0.0910	0.85	199
10	0.1411	11.67	20	0.5319	0.85	190	0.0919	0.83	196
15	0.1672	11.74	19	0.5346	0.98	190	0.1012	0.82	195
20	0.1750	11.75	18	0.5368	0.92	189	0.1156	0.82	192
25	0.1788	11.76	15	0.5484	0.90	187	0.1389	0.79	189
30	0.1932	12.68	14	0.5499	0.87	187	0.1596	0.78	189
35	0.2056	12.88	13	0.5506	0.93	186	0.1865	0.76	188
40	0.2806	12.88	13	0.5522	0.88	183	0.1954	0.73	186
45	0.2813	12.93	12	0.5723	0.98	180	0.2289	0.72	185
50	0.3030	14.25	11	0.6424	0.81	171	0.2467	0.71	179

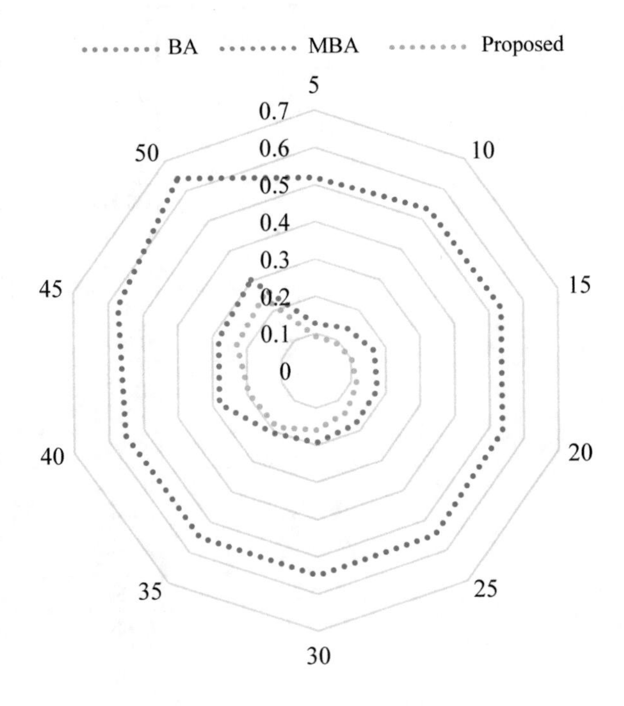

Fig. 9 Impact of ranging error in terms of MLE

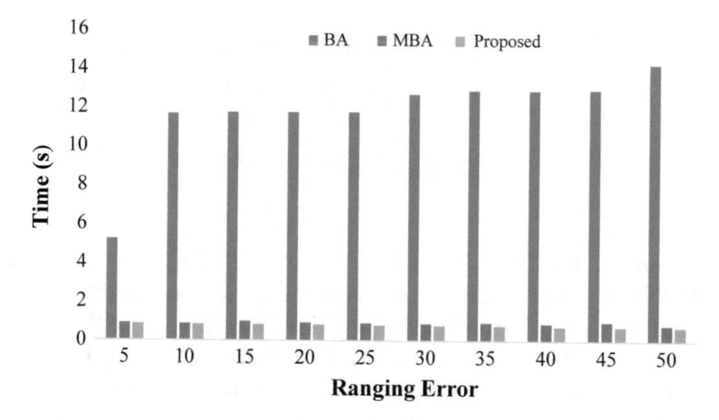

Fig. 10 Impact of ranging error in terms of computation time

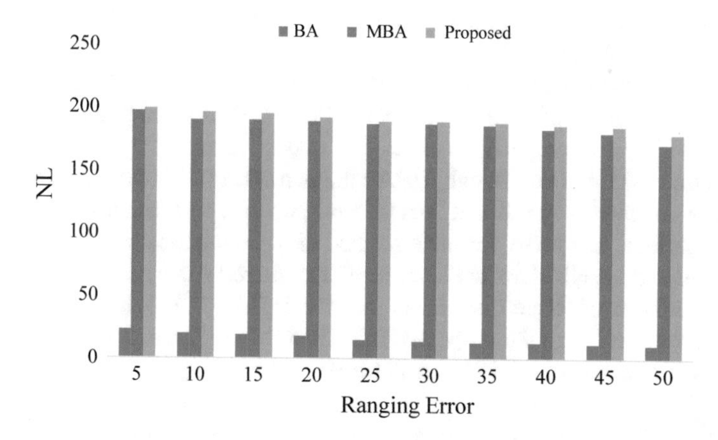

Fig. 11 Impact of ranging error in terms of NL

Firstly, Fig. 9 shows the performance analysis of different localization approaches in terms of MLE. From the figure, it is clearly shown that the MLE starts to increase with an increase in error rate. For the 5% error, the MLE of the BA, MBA and BAT-TLBO are 0.1260, 0.5191 and 0.0910 respectively. From these values, it is apparent that the BAT-TLBO achieves lowest MLE than the other methods. In addition, for the 50% error, the MLE BA, MBA and BAT-TLBO are 0.303, 0.6424 and 0.2467 respectively. These values imply that the MLE is increased from 5 to 50% and the addition of noise worsen the localization performance. Here, the MBA showed poor performance compared to BA and proposed algorithm. Compared to all the algorithms, the BAT-TLBO achieved lowest MLE on all the applied error rate.

Figure 10 illustrate the results of the localization algorithms on the influence of ranging error between 5 and 50% in terms of computation time. The values shown in figure ensure that the computation time is very high for the BA. Then, the BAT-TLBO consumes only less time compared to other algorithm on all the applied ranging error.

Table 5 Mean and standard deviation

Algorithm	Min.	Max.	Av.	SD.
BA	0.1215	0.4429	0.220986	0.064927
MBA	0.4753	0.6188	0.550076	0.027771
Proposed	0.1098	0.2789	0.197556	0.036057

The MBA shows competitive performance over BAT-TLBO algorithm. However, the BAT-TLBO outperforms the MBA. For instance, for the ranging error 5%, the computation time of the BA, MBA and BAT-TLBO are 5.22 s, 0.89 s and 0.85 s respectively. From these values, it is apparent that the BAT-TLBO achieves lowest MLE than the other methods. From the values, it is revealed that the computation time also increase with increasing error rate. This can be proved by comparing the computation time needed by the localization algorithm at the error rate of 5 and 50%. For the 50% error, the computation time of BA, MBA and BAT-TLBO are 14.25 s, 0.81 s and 0.71 s respectively.

Finally, the effectiveness of the localization algorithms area analyzed on the varying error rate in terms of the NL is shown in Fig. 11. As seen in figure, for 5% error rate, the BAT-TLBO properly localizes 199 nodes whereas the BA localizes only a minimum of 23 nodes. Though MBA attains a NL of 197 which is higher than the results obtained by the BA, it fails to show superior performance over the BAT-TLBO algorithm. Likewise, for the error rate of 50%, the localization performance is degraded and the NL starts to decrease with an increase in error rate. The order of increased number of NL of different algorithms is BAT-TLBO, MBA, and BA. For the 50% error, the NL of the proposed BAT-TLBO algorithm is 179 whereas the NL of the BA and MBA are only 11 and 171 nodes respectively.

Table 5 provides the mean and standard deviation values of the proposed BAT-TLBO algorithm compared to BA and MBA. From these values, it is depicted that the proposed algorithm attains a minimum MLE of 0.1098 whereas the existing BA and MBA are 0.4753 and 0.1098 respectively. Meanwhile, the proposed BAT-TLBO algorithm achieves a maximum of only 0.2789 whereas the MBA obtains a value of 0.6188. This implies that the proposed algorithm is found to be efficient than the compared algorithms.

5 Conclusion

This paper has presented a new node localization algorithm for MWSN by integrating the bat algorithm with the TLBO algorithm. The proposed BAT-TLBO algorithm employs the nature of bat's echolocation and the characteristics of TLBO algorithm. An extensive set of experiments takes place to validate the results of the BAT-TLBO algorithm to resolve the node localization issue. To verify the consistency in the performance, the proposed and compared BA and MBA algorithms are tested with varying anchor node density and ranging error. From the experimental values, it is ensured that the BAT-TLBO algorithm effectively localizes more number of nodes with less computation time and MLE than the state of art methods. In future, the BAT-TLBO can be improved to reduce the computation time take to complete the localization process in MWSN.

References

1. M. Elhoseny, A.E. Hassanien, Mobile object tracking in wide environments using WSNs, in *Dynamic Wireless Sensor Networks. Studies in Systems, Decision and Control*, vol. 165 (Springer, Cham, 2019), pp 3–28. https://doi.org/10.1007/978-3-319-92807-4_1
2. H. Karl, A. Willig, *Protocols and Architectures for Wireless Sensor Networks* (Wiley, New York, 2005), p. 526
3. I.F. Akyildiz, W. Su, Y. Sankarasubramaniam, E. Cayirci, A survey on sensor networks. IEEE Commun. Mag. **40**(8), 102–114 (2002)
4. L. Chelouah, F. Semchedine, L. Bouallouche-Medjkoune, Localization protocols for mobile wireless sensor networks: a survey. Comput. Electr. Eng. 1–19 (2017). https://doi.org/10.1016/j.compeleceng.2017.03.024
5. W. Elsayed, M. Elhoseny, S. Sabbeh, A. Riad, Self-maintenance model for wireless sensor networks. Comput. Electr. Eng. (2017). https://doi.org/10.1016/j.compeleceng.2017.12.022
6. N. Sabor, S. Sasaki, M. Abo-Zahhad, S.M. Ahmed, A comprehensive survey on hierarchical-based routing protocols for mobile wireless sensor networks: review, taxonomy, and future directions. Wirel. Commun. Mob. Comput. **2017**, 23, Article ID 2818542 (2017). https://doi.org/10.1155/2017/2818542
7. A. Pal, Localization algorithm in wireless sensor networks: current approaches and future challenges. Netw. Protocols Algorithms **2**(1), 45–73 (2010)
8. A. Kannan, G. Mao, B. Vucetic, Simulated annealing based wireless sensor network localization. J. Comput. **1**(2), 15–22 (2006)
9. P.S. Raman, K. Shankar, M. Ilayaraja, Securing cluster based routing against cooperative black hole attack in mobile ad hoc network. Int. J. Eng. Technol. **7**(9), 6–9 (2018)
10. L. Doherty, Convex position estimation in wireless sensor networks, in *Twentieth Annual Joint Conference of the IEEE Computer and Communication Societies* (2001)
11. M. Elhoseny, A. Farouk, N. Zhou, M.-M. Wang, S. Abdalla, J. Batle, Dynamic multi-hop clustering in a wireless sensor network: performance improvement. Wirel. Pers. Commun. (Springer US) **95**(4), 3733–3753. https://doi.org/10.1007/s11277-017-4023-8
12. R. Kulkarni, G. Venayagamoorthy, M. Cheng, Bio-inspired node localization in wireless sensor networks, in *Proceedings of IEEE International Conference on Systems, Man, and Cybernetics*, New York (2009)
13. X. Yuan, M. Elhoseny, H.K. El-Minir, A.M. Riad, A genetic algorithm-based, dynamic clustering method towards improved WSN longevity. J. Netw. Syst. Manag. (Springer US) **25**(1), 21–46 (2017). https://doi.org/10.1007/s10922-016-9379-7
14. M. Elhoseny, X. Yuan, H.K. ElMinir, A.M. Riad, An energy efficient encryption method for secure dynamic WSN. Secur. Commun. Netw. (Wiley) **9**(13), 2024–2031 (2016). https://doi.org/10.1002/sec.1459
15. M. Elhoseny, H. Elminir, A. Riad, X. Yuan, A secure data routing schema for WSN using Elliptic Curve Cryptography and homomorphic encryption. J. King Saud Univ. Comput. Inf. Sci. (Elsevier) **28**(3), 262–275 (2016). http://dx.doi.org/10.1016/j.jksuci.2015.11.001
16. M. Elhoseny, X. Yuan, Z. Yu, C. Mao, H. El-Minir, A. Riad, Balancing energy consumption in heterogeneous wireless sensor networks using genetic algorithm. IEEE Commun. Lett. **19**(12), 2194–2197 (2015). https://doi.org/10.1109/LCOMM.2014.2381226

17. M. Elhoseny, K. Shankar, S.K. Lakshmanaprabu, A. Maseleno, N. Arunkumar, Hybrid optimization with cryptography encryption for medical image security in Internet of Things. Neural Comput. Appl. (2018). https://doi.org/10.1007/s00521-018-3801-x
18. K. Shankar, S.K. Lakshmanaprabu, D. Gupta, A. Maseleno, V.H.C. de Albuquerque, Optimal feature-based multi-kernel SVM approach for thyroid disease classification. J. Supercomput. (2018). https://doi.org/10.1007/s11227-018-2469-4
19. S.K. Lakshmanaprabu, K. Shankar, D. Gupta, A. Khanna, J.J.P.C. Rodrigues, P.R. Pinheiro, V.H.C. de Albuquerque, Ranking analysis for online customer reviews of products using opinion mining with clustering. Complexity **2018**, Article ID 3569351, 9 (2018). https://doi.org/10.1155/2018/3569351
20. S.K. Lakshmanaprabu, K. Shankar, A. Khanna, D. Gupta, J.J. Rodrigues, P.R. Pinheiro, V.H.C. De Albuquerque, Effective features to classify Big Data using social Internet of Things. IEEE Access **6**, 24196–24204 (2018)
21. Q. Zhang, J. Wang, C. Jin, Q. Zeng, Localization algorithm for wireless sensor networks based on genetic simulated annealing algorithm, in *International Conference on Wireless Communication Networking and Mobile Computing* (2008)
22. Q. Zhang, J. Huang, J. Wang, C. Jin, J. Ye, W. Zhang, A new centralized localization algorithm for wireless sensor network, in *Third International Conference on Communications and Networking in China, Physical World with Pervasive Networks* (2008)
23. Y. Li, J. Xing, Q. Yang, H. Shi, Localization research based on improved simulated annealing algorithm in WSN, in *5th International Communications of Conference on Wireless Communications, Networking and Mobile Computing* (2009)
24. A. Gopakumar, L. Jacob, Localization in wireless sensor networks using particle swarm optimization, in *IET International Conference on Wireless, Mobile Multimedia Networks*, New York (2008)
25. Q.S. Zhao, G.Y. Meng, A multidimensional scaling localisation algorithm based on bacterial foraging algorithm. Int. J. Wirel. Mob. Comput. **6**(1), 58–65 (2013)
26. R.V. Rao, V.J. Savsani, D.P. Vakharia, Teaching–learning-based optimization: a novel method for constrained mechanical design optimization problems. Comput. Aided Des. **43**(3), 303–315 (2011)
27. K. Karthikeyan, R. Sunder, K. Shankar, S.K. Lakshmanaprabu, V. Vijayakumar, M. Elhoseny, G. Manogaran, Energy consumption analysis of Virtual Machine migration in cloud using hybrid swarm optimization (ABC–BA). J. Supercomput. (2018). https://doi.org/10.1007/s11227-018-2583-3
28. D.R. Griffin, F.A. Webster, C.R. Michael, The echolocation of flying insects by bats. Anim. Behav. **8**(34), 141–154 (1960)
29. W. Metzner, Echolocation behaviour in bats. Sci. Prog. Edinb. **75**(298), 453–465 (1991)
30. H.-U. Schnitzler, E.K.V. Kalko, Echolocation by insect-eating bats. Bioscience **51**(7), 557–569 (2001)
31. X.-S. Yang, Bat algorithm for multi-objective optimisation. Int. J. Bio-Inspir. Comput. **3**(5), 267–274 (2011)
32. M. Elhoseny, A. Tharwat, A. Farouk, A.E. Hassanien, K-coverage model based on genetic algorithm to extend WSN lifetime. IEEE Sens. Lett. **1**(4), 1–4 (2017). https://doi.org/10.1109/lsens.2017.2724846

A Next Generation Hybrid Scheme Mobile Graphical Authenticator

Teoh Joo Fong, Azween Abdullah and Hamid Reza Boveiri

Abstract **Objectives**: To provide a swift and simple mobile authentication method which are highly secured, easily remembered and prevents shoulder surfing attacks to improve existing mobile authentication methods. **Methods**: This paper is written using a problem-oriented research in improving the existing mobile authenticator which are vulnerable to shoulder surfing attack. Several qualitative researches are done by analyzing other related work done in the graphical authenticator field which are solving the same problem. A quantitative experiment method is used to test the proposed solution. **Findings**: Currently, most mobile devices are protected by a six pins numerical passcode authentication layer which is extremely vulnerable to Shoulder Surfing attacks and Spyware attacks. This paper proposes a multi-elemental graphical password authentication model for mobile devices that are resistant to shoulder surfing attacks and spyware attacks. The proposed Coin Passcode model simplifies the complex user interface issues that previous graphical password models have, which work as a swift passcode security mechanism for mobile devices. The Coin Passcode model also has a high memorability rate compared to the existing numerical and alphanumerical passwords, as psychology studies suggest that humans are better at remembering graphics than words. **Novelty**: Implementing multiple hidden elements in one button passcode which shuffles randomly to prevent shoulder surfing attack in mobile authenticator.

Keywords Mobile graphical password · Multi-elemental passcode · Shoulder-surfing proof passcode · Mobile authentication model

Both A. Abdullah and H. R. Boveiri are identically the corresponding authors.

T. J. Fong · A. Abdullah (✉)
School of Computing and IT, Taylor's University, Subang Jaya, Malaysia
e-mail: azween.abdullah@taylors.edu.my

H. R. Boveiri (✉)
Sama Technical and Vocational Training College, Islamic Azad University, Shoushtar Branch, Shoushtar, Iran
e-mail: boveiri@samashoushtar.ac.ir

M. Elhoseny and A. K. Singh (eds.), *Smart Network Inspired Paradigm and Approaches in IoT Applications*, https://doi.org/10.1007/978-981-13-8614-5_14

1 Introduction

Mobile user authentication technology is crucial to the integrity and confidentiality of smart devices, especially when many vital features such as banking and finance are currently accessible through mobile applications. Current mobile security mechanisms use the four or six pin numerical passcodes which are easily remembered, while providing a swift security authentication for the users. However, this security mechanism has its flaws when it faces modern attackers who can easily guess or shoulder surf for the password combinations. There are several other user authentication mechanisms such as the alphanumerical passwords and the pattern drawing lock, which are also prone to shoulder surfing attacks. The two-factor authentication can be easily compromised when the first level protection of the mobile devices is vulnerable.

The Coin Passcode Graphical Password Authentication mechanism is a concept of the Cognitive biometric authentication which uses the hybrid scheme graphical password authentication mechanism. This paper is structured in six sections, including the Introduction Section, Related Works, The Coin Passcode Mobile Graphic Authentication Model, Security Analysis and Usability Metrics, Discussion and Conclusion.

1.1 Cognitive Biometrics

There are several different biometric authentication types including physiological biometrics and cognitive biometrics authentication. Under the Cognitive biometric authentication methods, user behaviors are identified using mobile phone sensors, through systems such as gestures and walking patterns. The input patterns are used as a means of behavior authentication. Several researchers have analyzed the key input mechanics and patterns used by the users when they press the on-screen buttons to type on the phone [1]. Another research conducted by Giuffrida et al. [2], allows Cognitive Authentication when the user types the password, where the movement and touch of the screen are analyzed and authenticated. According to Stanciu et al. [3], this method is effective enough to protect the system from statistic attack. An improved version of this method of authentication uses the combination of four aspects which are time, pressure, size and acceleration obtained from the sensor of the device when a user makes password input [4].

Other than Cognitive authentication via password input, there is also authentication done using pattern drawings [5]. Recognition of shape drawing patterns to authenticate a user is a strong and easy way to protect the user against password peeking attacks [6]. Another kind of Cognitive authentication method is through the user's pattern of walking [7]. Through this method, users wear a movement tracker attached to the waist to track the user's continuous movement data [8]. Besides that, a type of Cognitive biometrics authentication measures the usage pattern and geographical location of regular usage of a user with his smartphone. This method measures the user-phone interaction activity such as the application usage, location, communication and motion to detect anomaly intruder usage scenarios [9]. It is suitable for mobile devices to implement a Cognitive biometric authentication such as the graphical password authentication as there are external costs involved in purchasing devices with sensors. Graphical Password Authentication is an authentication method in the Cognitive Biometric Authentication category which is suitable for implementation in mobile devices compared to existing numerical passcodes.

Passwords in the form of graphics are secure alternatives to numerical and text-based passwords, where the users are required to select pictures for authentication instead of keying in texts [10]. Passwords in graphical format are much easier to remember compared to text-based passwords. Some studies of psychology have identified that the human brain is way better in memorizing and recognizing visualized information such as pictures, compared to information in the form of text or speech [11]. Pictures are increasingly used for the purpose of security compared to mere texts, as the range of texts and numbers is limited in comparison to pictures which are infinite.

1.2 Recognition-Based Techniques

One of the graphical password authentication technique is called the Recognition-Based Technique. For this technique, symbols, icons or images are selected by the users in a series as a password during registration, where the users have to identify the same pictures they have selected during the authentication period [12]. Dhamija and Perrig [13], introduced a method of authentication using predefined images. Through this method, users are required to select their pre-selected pictures which they have defined during their registration from a set of random images to get authenticated by the system. However, this method is vulnerable to shoulder surfing attack.

Another example of recognition-based graphical authentication is called Passface™. This technique will display nine faces on the screen and require the user to choose their pre-selected faces in four rounds, choosing one pre-selected face per round [14]. In addressing the shoulder surfing issue with a graphical recognition authentication method, Gao et al. [15] introduced a shoulder surfing prevention method using invisible pattern drawing by swiping gestures to select a sequence of predefined images instead of tapping. An image chain in a story is used to remember the picture sequences to provide the user with the authentication. This method is less likely to be considered as it is vulnerable to shoulder surfing attacks as it is considerably easier to be identified compared to numbers.

1.3 Recall-Based Techniques

Another technique for Graphical Password Authentication is the Recall-Based Technique, which is based on pure recall and requires the users to recreate the graphics without any given tips or assisting reminders. However, users may find it hard to recall their password with this technique even though it is more secure than recognition-based technique. A technique called Syukri, by Eilejtlawi [16] requires the user to make a signature with a drawing with a stylus pen or mouse during registration, and authentication will be based on the same signature drawing.

A similar technique based on recall is enhanced with cues, where users are required to recreate a graphical password with the assistance of tips to enhance the accuracy of the password, where images will be provided to the users in which they must select specific points in the pictures in the right sequence. An example of this technique is a method introduced by Chiasson et al. [17], where the next picture on sequence is shown depending on the point of the previous click by the user. Every picture shown next to the previous picture based on a coordinate function of the point of click by the user of the current picture. A wrong selection of a point will cause the next picture to be shown wrongly, which prevents the attacker from guessing the password without knowing the right point for clicking.

1.4 Hybrid Scheme

A combination of multiple graphical password authentication method forms a hybrid scheme. The hybrid scheme is proposed by researchers in addressing the issues with limitation in every graphical authentication technique like shoulder surfing attacks, hotspot issues, and much more. Zhao and Li [18] introduced an example of a hybrid scheme, which is a text-in-graphic password authentication scheme—S3PAS in short, to counter shoulder surfing attacks.

A combination of texts and graphics can resist shoulder surfing attacks, hidden cameras and spyware. The method of registration requires the user to choose "k", an original string text password. In the login authentication, the user has to look for the pre-defined password in the image, which will form an invisible triangle named the "passtriangle", and the user must then click in the region inside the invisible triangle to gain access. ChoCD is a hybrid graphical authentication system called ChoCD proposed by Radhi and Mohd [19]. ChoCD is a system which allows the user to sign in with a User ID and a graphical password. The system is implementable in both desktop and smart phones. The system authenticates a user in three ways, from the first step based on choice selection to the second step based on clicks and thirdly based on drawings. The scheme only allows the authenticated user to be able to recognize the passwords through graphic, clicking positions and drawing patterns. Users should be able to remember the pattern when the images are shown.

2 The Coin Passcode Graphical Password Authentication Model

The Coin Passcode Graphical Password Authentication Model is a hybrid graphical password mobile authentication scheme. It is relatively swift to be inputted as a mobile device authentication mechanism compared to a four or six pins numerical passcode. The key feature of the Coin Passcode Model is its resistance against shoulder surfing attacks and brute-force attacks for mobile devices.

The identity verification of a smart device user will be done through the validation of a set of Coin Passcode Graphical Passwords Keypair Authentication process, where the user will initially register a set of Coin Passcode graphical password patterns to be remembered, and by inputting the correct sequence of coin passcode, patterned graphical passwords would authenticate the identity of the user by means of cognitive biometric password. This can prevent an unauthorized user from getting access to the mobile device.

2.1 The Coin Passcode Structure

The Coin Model Graphical Password Authentication uses the concept of multi-elements found in the structure of any currency coin. In coins from different countries, there is always a combination of different symbols, numerical values, and some wordings. As with the concept of coins, the Coin Model Graphical Password Authentication uses the element of colors, numerical values, and icons to form unique coin passcodes as shown in Fig. 1. The color codes are added in the Coin to assist color-blind users.

There are a total of 10 icons, ten numbers and ten colors used as the elements of the Coin Passcode. The list of the element items is illustrated in Fig. 2. The icons are obtained from Google Material Icons for Android Development.

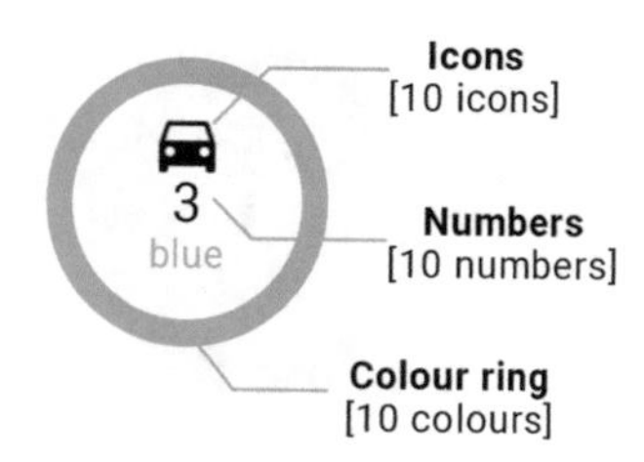

Fig. 1 The Coin Passcode element structure

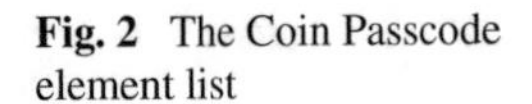
Fig. 2 The Coin Passcode element list

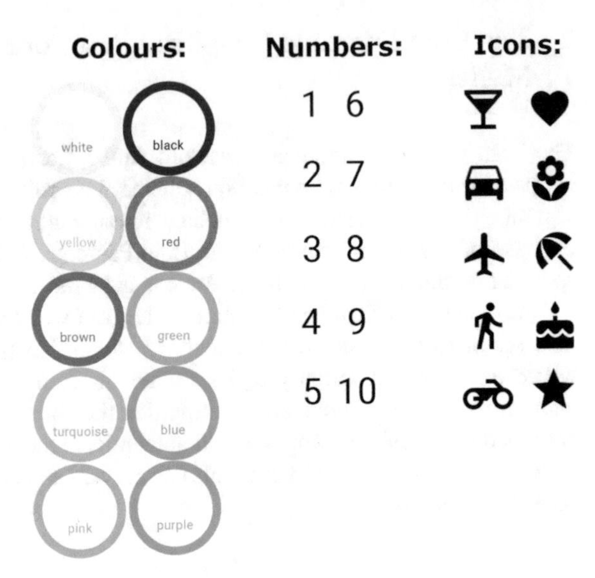

2.2 *The Coin Passcode Keypad Randomization*

The elements in the Coin Password are randomized each time, where every Coin Password will have a unique and different set of elements consisting of colors, numbers, and icons. There are a total of ten Coin Passwords in each different input attempt. With each Coin Password selected, a new layer will be formed, showing another randomized set of ten Coin Passwords, until the password authentication matches. An example of randomized Coin Passcode is shown in Fig. 3, in which each attempt shows the number 3 with different colors and icon elements.

The randomization algorithm pseudocode of the keypad elements is shown in Fig. 4. The algorithm loop through 6 times to randomize the keypad elements 6 times for all 6 inputs.

Fig. 3 Example of Coin Passcode elements randomization

```
procedure randomizeGraphicSelections

    loop 6 times

        Declare array of coinPasswordObject = []

        Retrieve available option values from data store
        place colour options to colourArray
        place number options to numberArray
        place icon options to icon iconArray


        loop 10 times
            generate random number from 0 to length of colour array

            generateCoinPasswordObjectFromRandomNumber(randomNumber, colourArray, numberArray, iconArray)
            removeSelectedOptionItem(index, colourArray, numberArray, iconArray)

            add generateCoinPasswordObject to array of coinPasswordObject
        end loop

        Display values of coinPasswordObject array randomly

    end loop

end procedure
```

Fig. 4 The keypad elements randomization pseudocode

2.3 The Coin Passcode Registration

To strengthen the complexity of the Coin Password, a minimum password standard is set. Each of the three elements must be present in the Coin Password Combination at least twice, resulting in a combination of six coins in a passcode sequence with all three different hidden elements.

The Coin Passcode limits the user to place precisely six Coin Passcodes elements in the sequence during registration. The registering of the Coin Passcodes requires one secret hidden element item from each sequence to be initialized by the user during registration. For example, if the hidden element of the first Coin Password chosen is the color element 'Yellow', then the color 'Yellow' would be the key to the first Coin Password, ignoring the other two elements in the first Coin Password, which are the randomized numbers and icons as shown in Fig. 5.

The registration algorithm limits the user to select all three different element items in the first three coin passcodes element selection by removing an element to be selected after each selection. The algorithm then loops again for the last three coin passcodes selections with the same limitation as shown in Fig. 6.

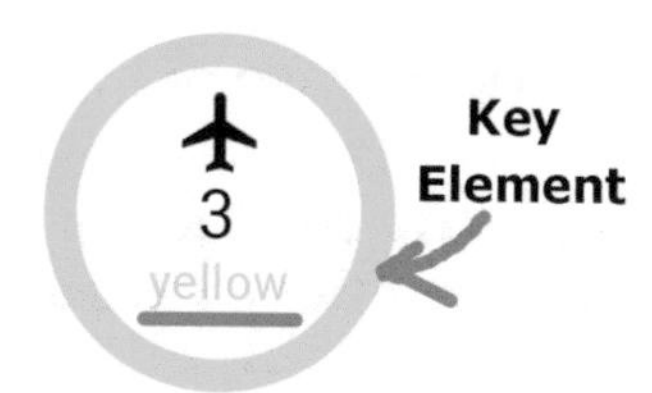

Fig. 5 The key element example in Coin Passcode

```
procedure coinPasswordRegistration

    declare coinPasswordObjectArray

    loop 6 times

        declare primaryElementsToExcludeArray = []
        declare coinPasswordObject

        loop 3 times
            displayPrimaryElementsWithExclusion(primaryElementsArray, primaryElementsToExcludeArray)
            User selects an primary graphical element
            add selected primary graphical element to primaryElementsToExcludeArray

            displayElements(selectedPrimaryElement)
            choose displayed options

            addCoinObjectProperty(coinPasswordObject, selectedOption)

        end loop

        add coinPasswordObject to coinPasswordObjectArray

    end loop

    store coinPasswordObjectArray to datastore

end procedure
```

Fig. 6 Coin Passcode registration algorithm pseudocode

2.4 The Coin Passcode Authentication Algorithm

The Coin Graphical Password Authentication is designed in a way that only the authorized user knows the hidden element he or she registers out of the three elements in each coin, whether it's the color, number, or the icon. An example of a Coin Passcode Registered Sequence combination is shown in Fig. 7, with the first secret coin numerical element of "Three", a second coin with the secret color element "Yellow", and the third coin with the secret icon element "Car", continuing the rest of the three secret Coin Passcodes elements with the number "Five", the color "Blue", and finally the icon "Flower". The authentication system will then match the Coin Passcode inputs based on the registered Coin Passcode elements and sequence while ignoring the rest of the public elements in each of the six Coin Passcode inputs. Any other attempt by selecting coins without the elements in the right sequence will result in a failure in the identity authentication.

An object array is used to store the user's coin passcode login input to be matched with the registered coin passcode object array to check whether the login input object array contains the registered credentials in the right sequence for authentication.

3 Security Analysis and Usability Metrics

Security and Usability seem to be at odds. It is a view of some that improving one affects the other in a negative way [20]. Several experiments are conducted with a group of 50 students to carry out the security analysis and usability metrics for the Coin Passcode against other similar mobile authentication models including the Numerical Passcode, Alphanumeric Passwords, and Passfaces™. The experiments

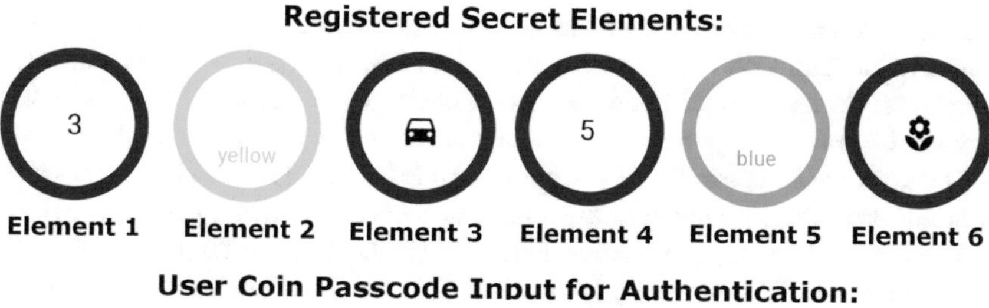

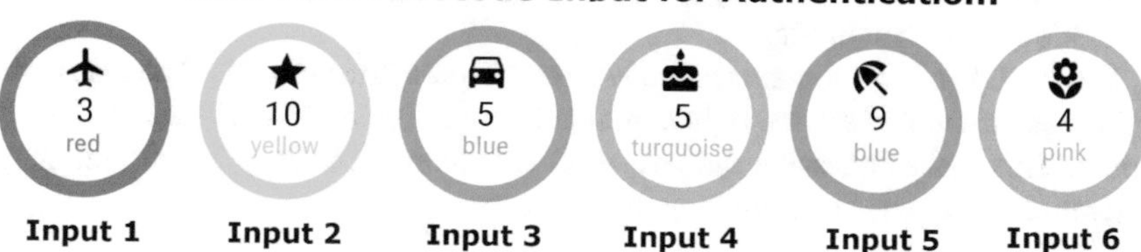

Fig. 7 Registered secret elements and authentication input example

Table 1 Password attack comparison table

Name of attacks	Password schemes			
	Numerical Passcode	Alphanumerical Password	Passfaces™	Coin Passcode
Shoulder surfing attack	Y	Y	Y	N
Dictionary attack	N	Y	N	N
Password guessing	Y	Y	Y	Y
Brute-force attack	Y	Y	Y	N
Spy ware attack	Y	Y	Y	N

conducted covers the usability metrics of login time and password memorability, and security analysis of shoulder surfing attack, password guessing attack and brute-force attack for each of the mentioned authentication models. Table 1 summarizes the security of the different password schemes against several password attack methods. "Y" refers to Yes, which it is vulnerable to the forms of attack. While "N" means No, which the password scheme is secured against the attack type.

3.1 *Password Complexity Comparison*

Passwords have been the focal point of many studies in the recent years. These studies have explored a range of related topics including password cracking algorithms [21–27], password strength and complexity [28–33], user behavior with respect to password selection [34–36], and password creation policies [37–41]. The Coin Passcode Authentication Model which consists of three elements in each coin creates a cognitive authentication link between the user and the authentication system, where only the right user would know the secret element and sequence he or she sets, leaving the rest of the people confused about the password. Based on the calculations below, the complexity of the Coin Passcode Model (1) is much more resistant to brute-force attacks compared to Numerical Passcodes (2) and Passfaces (4), but weaker compared to Alphanumerical Password (3) due to the differences in the number of elements. The complexity comparison chart of Coin Passcode and Numerical Passcode can be seen in Fig. 8.

It takes 729 million attempts to brute-force a triple elemental Coin Passcode in the right sequence to find out the right combination of the Coin Passcode secret element values. This makes guessing the password way more difficult compared the huge difference in passcode combination possibilities.

$(N \times E)^{L} =$ ***No. of Possible Passcode Combinations,***

$N =$ *No. of Input Buttons,* $E =$ *No. of Elements,*

$L =$ *Length of Passcode*

$(10 \times 3)^{6} = 729{,}000{,}000$ Coin Passcode Combinations (1)

$(10 \times 1)^{6} = 1{,}000{,}000$ Numerical Passcode Combinations (2)

$(62 \times 1)^{6} = 56{,}800{,}235{,}584$ Alphanumerical Combinations (3)

$(9 \times 1)^{6} = 531{,}441$ Passfaces™ Password Combinations (4)

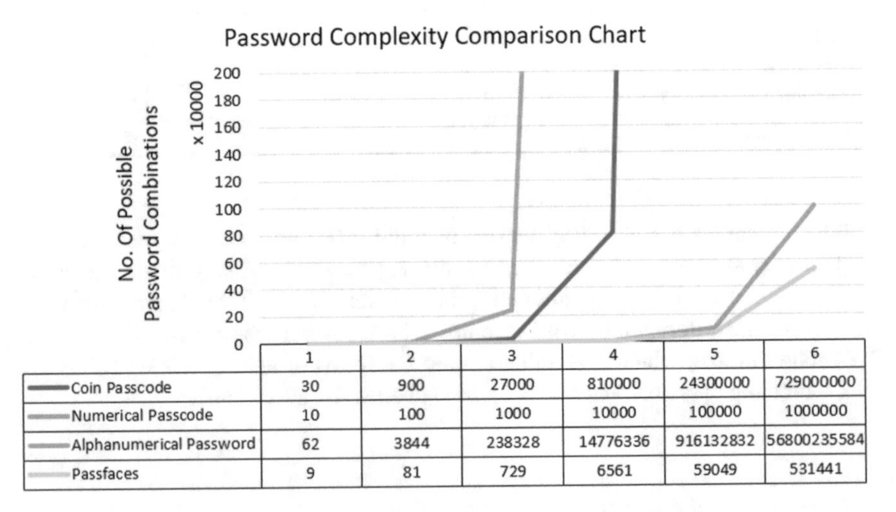

	1	2	3	4	5	6
Coin Passcode	30	900	27000	810000	24300000	729000000
Numerical Passcode	10	100	1000	10000	100000	1000000
Alphanumerical Password	62	3844	238328	14776336	916132832	56800235584
Passfaces	9	81	729	6561	59049	531441

Fig. 8 Password complexity comparison chart

3.2 *Shoulder Surfing Attack and Spyware Attack*

Shoulder Surfing attack uses the technique of direct observation or through recording using video cameras such as high-resolution surveillance equipment or hidden cameras to obtain a user's credentials. A spyware attack is when malwares are installed in a user's device to record the user's credentials input, while the information is sent back to the attacker for exploitation. Both of these attacks can easily obtain and exploit a user's numerical and alphanumerical password, or Passfaces™ credentials by directly observing the password input pressed by the user. However, the Coin Passcode is resistant to this type of attack.

An experiment is conducted with a group of 50 students, where a set of passwords for different authentication models, each with an equal password length of six items is pressed in front of the students through a big screen, with each button pressed at five-second intervals. The students are then asked to retype or reselect the shoulder surfed passwords. The result of the shoulder surfing attack experiment is shown in Fig. 9. The numerical passcode and alphanumerical password are seen to have a high rate of shoulder surfing success due to their vulnerability to this attack method. The Passfaces™, however has a lower success rate as it requires a certain recognition and memorability of the level of the faces used to reselect the right one.

The Coin Passcode can be observed to have zero success rate of shoulder surfing attack. This is because having multiple graphical elements in each input of the Coin Passcode would make shoulder surfing attack and spyware attack meaningless, as students do not get any direct password information from just observing the Coin Passcode combination inserted by the user. It is designed so that it is impossible for a shoulder surfer to know which secret element out of the three was the one being chosen by the user in a single input button.

Memorability is the measurement of the extent to which the users can remember the password after a period. A password memorability experiment is conducted

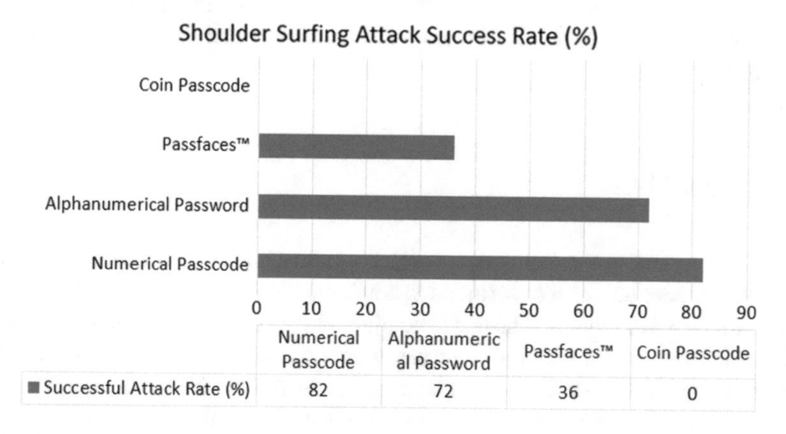

Fig. 9 Shoulder surfing attack success rate chart

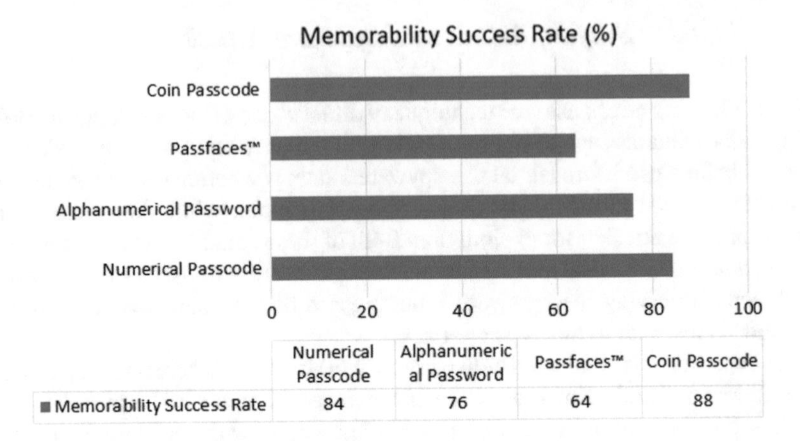

Fig. 10 Password memorability success rate chart

for each of the four password authentication models. The test is conducted with a group of 50 students, where each student is given a similar set of passwords of the same password length of six for each password model. The students were given five minutes to memorize each password and were then shown a 3 min video to simulate an extended period of idle time. After the video ended, the students were asked to produce the same password in one minute. The result of the experiment is shown in Fig. 10. The experiment result shows that the Coin Passcode has the highest memorability success rate followed by the numerical passcode, the alphanumerical password and lastly, the Passfaces™.

Based on the experiment, it was much easier to remember the Coin Passcode because the secret elements used are straightforward elements like colors, numbers, and icons, which can form a story-like chain of keywords such as "3 blue cars, 5 red bikes", as compared to remembering numbers, words or faces which have no direct meaning or connection to the tester. The experiment found that unfamiliar faces are hard to remember after a period of idle time, even though it is also a form of graphical password.

3.3 Login Time

Login Time refers to the time taken for users to log into the authentication system using their credentials. An experiment is conducted to analyze the login time for the four authentication models. The test is conducted with a group of 50 students, where each student is given a set of similar passwords of the same length of six for each password model. The students are then asked to reproduce the same passwords five times, and each login attempt time was recorded. The result of the experiment is shown in Fig. 11.

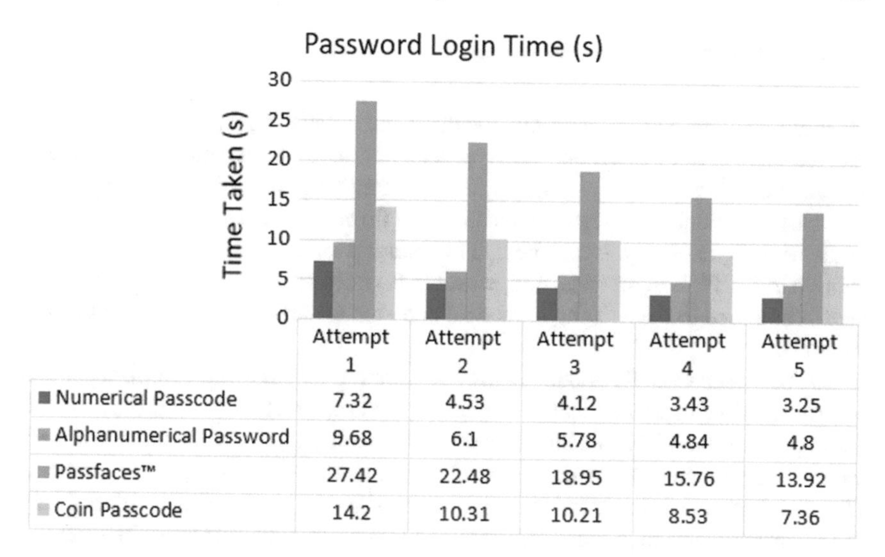

	Attempt 1	Attempt 2	Attempt 3	Attempt 4	Attempt 5
Numerical Passcode	7.32	4.53	4.12	3.43	3.25
Alphanumerical Password	9.68	6.1	5.78	4.84	4.8
Passfaces™	27.42	22.48	18.95	15.76	13.92
Coin Passcode	14.2	10.31	10.21	8.53	7.36

Fig. 11 Password login time chart

The Coin Passcode has slightly longer login time compared to Numerical Passcode and Alphanumerical Password of the same length even after five attempts. This is because the positioning of the numerical passcode and alphanumerical password are fixed, which the test user can simply memorize and get used to from each increased attempt. However, the positioning of the Coin Passcode elements is randomized and shuffled in each attempt, which is designed to confuse the shoulder surfing attacker, causing a longer login time for the test users. The Passfaces™ takes a much longer login time compared to the other password models due to the low memorability of the unfamiliar faces which requires the test user to take time to confirm the faces.

3.4 Password Guessing and Dictionary Attack

Password Guessing is a kind of brute-force attack which uses knowledge or hints gained from the password owner. Each of the password models mentioned in the analysis are vulnerable to this attack when the user leaves certain hints or information about their password exposed to the attacker. This attack cannot be avoided and can only be prevented through security awareness and training. A dictionary attack is conducted using a list of frequently used words or number patterns to crack the password efficiently. However, this only applies to the existing Numerical Passcodes and Alphanumerical Passwords due to the reason that these passwords often contain phrases that are predictable and highly used statistically. The Coin Passcode Authentication Model and the Passfaces™ does not apply in this case because a Dictionary Attack does not work for graphical passwords.

4 Discussions

Most graphical passwords currently available are mostly proven to be more secure and resistant to several cybersecurity attacks compared to existing numerical and alphanumerical passwords. However, these graphical passwords are mostly available only in the field of research, education and theoretical discussion, and are rarely implemented practically. It may be due to several poor usability factors such as low memorability, high login time, and non-user friendly or non-mobile friendly interfaces, compared to the existing numerical and alphabetical password authentication methods.

The proposed Coin Passcode is shown to have higher password complexity when compared to its closest identical numerical passcode model. Even though the alphanumeric password model has a higher password complexity, it is still not a completely secure password mechanism due to its vulnerability towards shoulder surfing attacks. The Coin Passcode is designed to overcome the shoulder surfing attack vulnerability and is currently designed specifically for a swift mobile authentication which greatly enhances the password complexity compared to its nearest comparison. A higher password complexity can be achieved when the coin passcode's multi-elemental concept is implemented to a similar input of alphanumerical passwords.

The memorability of the Coin Passcode is also a beneficial key feature due to its straightforward elemental attributes which can be formed into a chain of story-like keywords that other existing password models are missing. It is more likely for people to remember a story formed by visuals rather than numbers or alphabets. The login time for the Coin Passcode is not as fast as the existing password models due to the randomization and element shuffling nature of the Coin Passcode model. However, this can be considered a security over performance prioritization measure.

5 Conclusion

In conclusion, the Coin Passcode is able to overcome the current shoulder surfing and spyware attack vulnerability that existing mobile application numerical passcode authentication layers suffer from. It is shown that having a Multi-elemental passcode for a mobile login interface can prevent direct observation password attacks, and at the same time provide a higher password complexity against brute-force and password guessing attacks. It is a combination of the behavioral context uniqueness of each person that makes this multi-elemental passcode a stronger mobile password interface.

The authors believe in the real potential of graphical password in benefiting the current mobile smart devices swift authentication mechanism, in terms of usability and security aspects. This brings the purpose for us to propose the Coin Passcode Graphical Password Mobile Authentication Model in hoping to overcome the challenges by bringing a simplicity in usage plus complexity in security for the mobile developers as well as the mobile users. However, there are still limitations in the currently proposed design of the Coin Passcode which can be further enhanced in the future for the betterment of mobile security. One of it is the lack of encryption for the coin passcodes input and stored passcodes, as the elements are currently stored purely in plaintext and can be easily modified via code injection attack. The Multi-elemental input concept should also be further explored in password model fields other than mobile security layers, such as the network security and banking security.

References

1. N.L. Clarke, S.M. Furnell, Authenticating mobile phone users using keystroke analysis. Int. J. Inf. Secur. **6**(1), 1–14 (2007). Springer, New York
2. C. Giuffrida, K. Majdanik, M. Conti, H. Bos, I sensed it was you: authenticating mobile users with sensor-enhanced keystroke dynamics, in *International Conference on Detection of Intrusions and Malware, and Vulnerability Assessment*, ed. by L. Cavallaro (Springer, New York, 2014 July), pp. 92–111
3. V.D. Stanciu, R. Spolaor, M. Conti, C. Giuffrida, On the effectiveness of sensor-enhanced keystroke dynamics against statistical attacks, in *Proceedings of the Sixth ACM Conference on Data and Application Security and Privacy*, ed. by C. Busch, A. Brömme (ACM, New York, 2016 March), pp. 105–112
4. N. Zheng, K. Bai, H. Huang, H. Wang, You are how you touch: user verification on smartphones via tapping behaviors, in *2014 IEEE 22nd International Conference on Network Protocols*, ed. by J. Kaur, G. Rouskas (IEEE, New York, 2014 October), pp. 221–232
5. A. De Luca, A. Hang, F. Brudy, C. Lindner, H. Hussmann, Touch me once and I know it's you: implicit authentication based on touch screen patterns, in *Proceedings of the SIGCHI Conference on Human Factors in Computing Systems*, ed. by J.A. Konstan, E.H. Chi, Kristina Höök (ACM, New York, 2012 May), pp. 987–996
6. A. De Luca, M. Harbach, E. von Zezschwitz, M.E. Maurer, B.E. Slawik, H. Hussmann, M. Smith, Now you see me, now you don't: protecting smartphone authentication from shoulder surfers, in *Proceedings of the 32nd annual ACM conference on Human factors in computing systems*, ed. by M. Jones, P. Palanque (ACM, New York, 2014 April), pp. 2937–2946
7. J. Mantyjarvi, M. Lindholm, E. Vildjiounaite, S.M. Makela, H.A. Ailisto, Identifying users of portable devices from gait pattern with accelerometers, in *Proceedings (ICASSP'05). IEEE International Conference on Acoustics, Speech, and Signal Processing, 2005*, vol. 2, ed. by Petropulu (IEEE, New York, 2005 March), pp. ii–973
8. M.O. Derawi, C. Nickel, P. Bours, C. Busch, Unobtrusive user authentication on mobile phones using biometric gait recognition, in *Sixth International Conference on Intelligent Information Hiding and Multimedia Signal Processing (IIH-MSP)*, ed. by D.W. Fellner, X. Niu (IEEE, New York, 2010 October), pp. 306–311
9. E. Shi, Y. Niu, M. Jakobsson, R. Chow, Implicit authentication through learning user behavior, in *International Conference on Information Security*, ed. by S.K. Bandyopadhyay, W. Adi (Springer, Berlin, Heidelberg, 2010 October), pp. 99–113
10. L. De Angeli, G.J. Coventry, K. Renaud, Is a picture really worth a thousand words? Exploring the feasibility of graphical authentication systems. Int. J. Hum.-Comput. Stud. **63**(1–2), 128–152 (2005)
11. Kirkpatrick, An experimental study of memory. Psychol. Rev. **1**, 602–609 (1894)
12. K. Renaud, E. Smith, Jiminy: helping user to remember their passwords. Technical Report, School of Computing, University of South Africa (2001)
13. R. Dhamija, A. Perrig, Déjà Vu: a user study using images for authentication, in *9th USENIX Security Symposium* (2000)

14. T. Grinal, T. Aakriti, S. Akshata, R. Malvina, S. Aishwarya, Graphical password authentication using Pass faces. Int. J. Eng. Res. Appl. **5**(3), Part 5, 60–64 (2015 March)
15. H. Gao, X. Liu, R. Dai. Design and analysis of a graphical password scheme, in *International Conference on Innovative Computing, Information and Control (ICICIC)* (2009), pp. 675–678
16. A.M. Eilejtlawi, Study and development of a new graphical password system (2008 May)
17. S. Chiasson, P.C. van Oorschot, R. Biddle, Graphical password authentication using Cued Click Points, in *European Symposium on Research in Computer Security (ESORICS)*, LNCS 4734 (2007 September), pp. 359–374
18. H. Zhao, X. Li, S3PAS: a scalable shoulder-surfing resistant textual-graphical password authentication scheme, in *21st International Conference on Advanced Information Networking and Applications Workshops*, vol. 2 (Canada, 2007), pp. 467–472
19. R.A. Radhi, Z.J. Mohd, ChoCD: usable and secure graphical password authentication scheme. Indian J. Sci. Technol. **10**(4) (2017 January). 10.17485
20. K. Ronald, F. Ivan, A.W. Roscoe, *Security and Usability: Analysis and Evaluation* (Oxford University Computing Laboratory, 2010)
21. A. Narayanan, V. Shmatikov, Fast dictionary attacks on passwords using time-space tradeoff, in *Proceedings of the 12th ACM Conference on Computer and Communications Security, Series CCS '05* (ACM, New York, NY, USA, 2005), pp. 364–372
22. C. Castelluccia, C. Abdelberi, M. Durmuth, D. Perito, When privacy meets security: leveraging personal information for password cracking. CoRR, abs/1304.6584 (2013)
23. M. Weir, S. Aggarwal, B. de Medeiros, B. Glodek, Password cracking using probabilistic context-free grammars, in *Proceedings of the IEEE Symposium on Security and Privacy* (2009 May), pp. 391– 405
24. Z. Li, W. Han, W. Xu, A large-scale empirical analysis of chinese web passwords, in *Proceedings of 23rd USENIX Security Symposium, USENIX Security* (2014 August)
25. R. Veras, C. Collins, J. Thorpe, On the semantic patterns of passwords and their security impact, in *Proceedings of the Network and Distributed System Security Symposium (NDSS'14)* (2014)
26. J. Ma, W. Yang, M. Luo, N. Li, A study of probabilistic password models, in *Proceedings of the IEEE Symposium on Security and Privacy* (2014 May), pp. 689–704
27. B. Ur, S.M. Segreti, L. Bauer, N. Christin, L.F. Cranor, S. Komanduri, D. Kurilova, M.L. Mazurek, W. Melicher, R. Shay, Measuring real-world accuracies and biases in modeling password guessability, in *24th USENIX Security Symposium (USENIX Security 15)* (USENIX Association, Washington, D.C., 2015 August 2015), pp. 463–481
28. M. Dell'Amico, P. Michiardi, Y. Roudier, Password strength: an empirical analysis, in *INFOCOM, 2010 Proceedings IEEE* (2010 March), pp. 1–9
29. C. Castelluccia, M. Durmuth, D. Perito, Adaptive password-strength meters from markov models, in *Proceedings of the Network and Distributed System Security Symposium (NDSS)* (2012)
30. J. Bonneau, The science of guessing: analyzing an anonymized corpus of 70 million passwords, in *Proceedings of the IEEE Symposium on Security and Privacy* (2012 May), pp. 538–552
31. M.L. Mazurek, S. Komanduri, T. Vidas, L. Bauer, N. Christin, L.F. Cranor, P.G. Kelley, R. Shay, B. Ur, Measuring password guessability for an entire university, in *Proceedings of the 2013 ACM SIGSAC Conference on Computer & Communications Security, Series CCS '13* (ACM, New York, NY, USA, 2013), pp. 173–186
32. X. de Carne de Carnavalet, M. Mannan, From very weak to very strong: analyzing password-strength meters, in *Network and Distributed System Security (NDSS) Symposium 2014* (Internet Society, 2014 February)
33. Passfault, http://www.passfault.com/
34. D. Florencio, C. Herley, A large-scale study of web password habits, in *Proceedings of the 16th International Conference on World Wide Web, Series WWW '07* (ACM, New York, NY, USA, 2007), pp. 657–666
35. B. Ur, P.G. Kelley, S. Komanduri, J. Lee, M. Maass, M.L. Mazurek, T. Passaro, R. Shay, T. Vidas, L. Bauer et al., How does your password measure up? The effect of strength meters on password creation, in *USENIX Security Symposium* (2012), pp. 65–80

36. M. Weir, S. Aggarwal, M. Collins, H. Stern, Testing metrics for password creation policies by attacking large sets of revealed passwords, in *Proceedings of the 17th ACM Conference on Computer and Communications Security, Series CCS '10* (ACM, New York, NY, USA, 2010), pp. 162–175
37. R. Shay, S. Komanduri, P.G. Kelley, P.G. Leon, M.L. Mazurek, L. Bauer, N. Christin, L.F. Cranor, Encountering stronger password requirements: user attitudes and behaviors, in *Proceedings of the Sixth Symposium on Usable Privacy and Security, Series SOUPS '10* (ACM, New York, NY, USA, 2010), pp. 2:1–2:20
38. S. Komanduri, R. Shay, P.G. Kelley, M.L. Mazurek, L. Bauer, N. Christin, L.F. Cranor, S. Egelman, Of passwords and people: measuring the effect of password-composition policies, in *Proceedings of the SIGCHI Conference on Human Factors in Computing Systems, Series CHI '11* (ACM, New York, NY, USA, 2011), pp. 2595–2604
39. R. Shay, P.G. Kelley, S. Komanduri, M.L. Mazurek, B. Ur, T. Vidas, L. Bauer, N. Christin, L.F. Cranor, Correct horse battery staple: exploring the usability of system-assigned passphrases, in *Proceedings of the Eighth Symposium on Usable Privacy and Security, Seres SOUPS '12* (ACM, New York, NY, USA, 2012), pp. 1–20
40. P. Kelley, S. Komanduri, M. Mazurek, R. Shay, T. Vidas, L. Bauer, N. Christin, L. Cranor, J. Lopez, Guess again (and again and again): measuring password strength by simulating password cracking algorithms, in *2012 IEEE Symposium on Security and Privacy (SP)* (2012 May), pp. 523–537
41. R. Shay, S. Komanduri, A.L. Durity, P.S. Huh, M.L. Mazurek, S.M. Segreti, B. Ur, L. Bauer, N. Christin, L.F. Cranor, Can long passwords be secure and usable? in *Proceedings of the 32Nd Annual ACM Conference on Human Factors in Computing Systems, Seres CHI '14* (ACM, New York, NY, USA, 2014), pp. 2927–2936

Energy Efficient MANET by Trusted Node Identification Using IHSO Optimization

S. Krishnaveni and N. Angel

Abstract In Mobile Ad Hoc Network (MANET), the nodes are communicating with each and every node without any centralized admin or base station. Without any node awareness, it runs out of power in routing protocol which leads to major issues in energy efficiency and the node's link lifetime. To extend the energy efficiency of a routing protocol in MANET, proposed a protocol called Energy Efficient Ad hoc On-demand Distance Vectors Routing (EE-AODV). In the proposed study, the trusted nodes of MANET are identified for effective communication by the network parameter optimization. For optimizing the parameters, the random nodes are grouped by a new clustering model i.e. Virtual Link Weight based Clustering (VLWBC) and then trusted nodes are identified from each cluster by a metaheuristic algorithm i.e. Improved Harmony Search Optimization (IHSO). This IHSO algorithm optimizes the parameters such as QoS, energy consumption and network lifetime of MANET. The result demonstrates that the optimized MANET parameter improves the energy efficiency in all mobile nodes also and it achieves high QoS compared to existing algorithms.

Keywords Mobile Ad hoc network (MANET) · Routing protocol EE-AODV · VLWBC · Trusted node · IHSO · QoS

1 Introduction

Mobile ad hoc network is made out of mobile nodes that are speaking with every node without any administration or base station [1]. It is a consistently self-organizing, self-designing and infrastructure less network. It implies in this framework, nodes

S. Krishnaveni (✉)
Department of Computer Science, Pioneer College of Arts and Science, Coimbatore, India
e-mail: sss.veni@gmail.com

N. Angel
Department of Computer Science & Applications, Annai Veilankanni's College for Women, Chennai, India
e-mail: angel.nesaian@gmail.com

M. Elhoseny and A. K. Singh (eds.), *Smart Network Inspired Paradigm and Approaches in IoT Applications*, https://doi.org/10.1007/978-981-13-8614-5_15

are allowed to move in an arbitrary example [2]. Along these lines, every node acts as a router as well as a host. In MANET, routing is a procedure for choosing a path in the network [3]. It has two activities as—for finding an ideal routing way and to exchange data packet from source to destination way [4]. Network topology in MANET's is quickly changing as a result of the dynamic nature of nodes. In this way, routing protocols are utilized for dealing with the data packet between the courses and furthermore in charge of choosing the route in which data are conveyed to revise destination [5, 6]. It assumes an essential job in wireless communication. The mobile nodes in MANET are normally battery obliged which means energy efficient routing is an essential premise in outlining of these networks [7, 30].

There have been numerous MANET routing protocols, which fall into a few classifications: proactive routing protocols, for example, dynamic Destination-Sequenced Distance-Vector Routing (DSDV), Optimized Link State Routing (OLSR), Topology Broadcast in light of Reverse Path Forwarding (TBRPF), on-request routing protocols, for example, Dynamic Source Routing (DSR) [8, 31–33], Ad Hoc On-Demand Remove Vector (AODV) [9], Signal Stability based Adaptive Routing (SSA). Proactive routing protocols have little postponement for route disclosure and are sufficiently vigorous to interface breaks and get a worldwide ideal route for every destination [10, 34]. Be that as it may, their routing overhead is likewise high. On-request routing protocols are anything but difficult to acknowledge and their overhead is low [11]. Be that as it may, routes in on-request routing protocols are easy to break on account of topology varieties.

In traditional AODV routing, route request RREQ and route reply RREP control messages are utilized for route discovery process [12]. Ordinarily, a node broadcasts RREQ message to its neighboring node yet in at some point the middle of the road node will be unable to forward the RREP message on switch way in light of its inadequate energy [13, 14]. Consequently, the source node would need to rebroadcast the RREQ message in order to look a path for communication to the destination node [15]. This leads to increment in end-to-end delay of packets. In this paper, we have proposed an energy efficient routing convention which depends on existing AODV routing protocol [16]. It upgrades the battery lifetime of MANET node with no data or packet dropping in AODV [17] and the network parameters are enhanced by the innovative optimization algorithm. The principal target of this paper is to expand energy efficiency, network lifetime, and QoS and lessening the end-to-end deferral of the network [18]. It naturally increases the performance of AODV routing protocol and improves the energy of nodes in MANET. This work is organized as Sect. 2 discusses the existing papers related to WSN, Sect. 3 describes the routing mechanism in MANET, Sect. 4 describes the proposed network parameter optimization and mobile nodes clustering, Sect. 5 describes the results analysis of proposed VLWBC-IHSO model. At last, the conclusion part is described in Sect. 6.

2 Literature Review

In 2015 Chavhan and Venkataram [19] proposed a novel plan of giving QoS routing in MANETs by utilizing Emergent Intelligence (EI). The EI is a gathering knowledge, which is gotten from the periodical collaboration among a gathering of specialists and nodes. The mobile specialist connects with the nodes, neighboring mobile operators and static specialist for the accumulation of QoS asset data, finding secure and dependable nodes and finding an ideal QoS path from source to destination. The outcomes demonstrated that the effectiveness of the proposed work.

Venkatasubramanian and also Gopalan [20] clarified the enhanced energy-effective multipath QoS in MANET. By utilizing a QoS based Energy Efficient Optimal Multipath route discovery and route support in multipath routing convention, find the ideal way for data transmission from the source to the destination. Along these lines, QoS parameters, for example, transfer speed, throughput, packet delivery ratio and end-to-end delay will be figured and achieve in the most ideal way.

In 2016 Kuo and Chu [21] discussed the cross-layer design for MANETs by using Energy Efficient (EE) optimization. The problem is modeled as non-convex Mixed Integer Nonlinear Programming (MINLP) by routing, packet scheduling along with energy control. By using the branch and bound (BB) algorithm the defined problem is solved optimally. The results decrease the optimality gap 81.98% and increase the best feasible solution of 32.79% compared with the existing algorithm.

In 2013 Sridhar et al. [22] energy based EN-AODV protocol was suggested that distinguishes the nodes that deplete out of energy level at the time of information transmission. A node which has adequate energy level for the transmission is chosen for routing. The work was actualized and simulated on NS-2. The results have demonstrated an expansion in PDR, diminish in delay and throughput is kept up. The proposed EN-AODV gives more predictable as well as dependable data transfer contrasted with general AODV.

Mandhare et al. [23] concentrated on, fulfilling the imperative of QoS in MANET rousing Cuckoo Search (CS) optimization, in view of improving regular CS system utilizing on-demand protocol. This methodology chooses a QoS path based on the fitness value instead of a shortestway for Route Replay (RRPLY) packet of AODV. The performance of the CS optimization is analyzed in terms of mobility, adaptability, and congestion and proved CS is better than existing calculations.

In 2013 Jamali et al. [24] utilized the Binary Particle Swarm Optimization (BPSO) to add the energy awareness to the Temporally Ordered Routing Algorithm (TORA) routing protocol. The TORA protocol plans the routing issue as an enhancement issue and after that utilizes BPSO to pick a route that amplifies a weighted capacity of the route length and the route energy level. Using ns-2 tool demonstrated that the proposed routing BPSO-TORA draws out the network lifetime amazingly and add up to conveyed information.

3 Importance of Routing Mechanism in MANET

In MANETs, the nodes with high QoS based routing and keeping the link status up-to-date is a critical issue. Also, due to their dynamic characteristics, maintaining the precise link state information is very difficult. Hence, an enhanced AODV routing protocol is presented i.e. EE-AODV which is altered to improve the networks lifetime in MANET. By using this EE-AODV protocol, the best path is selected for data transmission with high residual energy and maximized network lifetime.

4 Methodology

A MANET is a gathering of wireless nodes that can be set up without any infrastructure or unified administrator. These nodes can act as in two ways for the purpose of forwarding the packet to different nodes, namely hosts as well as the router [37–40]. Routing the nodes with high QoS, restricted energy and so forth, is complex because of its dynamic topology. The methodology aims to improve QoS and to minimize the energy consumption by the optimization of nodes parameter and efficient routing. For clustering the random nodes and its head selection, VLWBC is introduced. And for energy efficient communication between the nodes, EE-AODV protocol is proposed which without disconnecting the route after link failure issue. From the clustered nodes, the optimal trusted nodes are identified by optimizing the network parameters like energy consumption, network lifetime and QoS using IHSO algorithm. Based on the optimized parameters, the trusted nodes are recognized for transmitting data from the source node to a destination with high energy efficient routing path. Figure 1 explains the overall process involved in the proposed work.

4.1 Routing Protocol Used in the Proposed Model

Routing is one of the significant difficulties in MANETs because of their exceptionally dynamic and distributed property. MANET routing protocols rely on how the protocols handle the packet to deliver from source to destination. Here, EE-AODV protocol is utilized.

EE-AODV: The proposed EE-AODV protocol gives energy efficient mechanism in MANET and in this protocol, the transmission power of the nearby nodes are reduced which in turn increased the battery lifetime of MANET nodes [25].

Route Establishment: In this stage, it will create the RREQ packet to find the best shortest way. It comprises of two procedures: Generate RREQ's and Processing and sending RREQ's. EE-AODV predicts the node quality by looking at the present threshold estimation of Received Signal Strength (RSS) with the received signal, choose whether this node will act as sending or not.

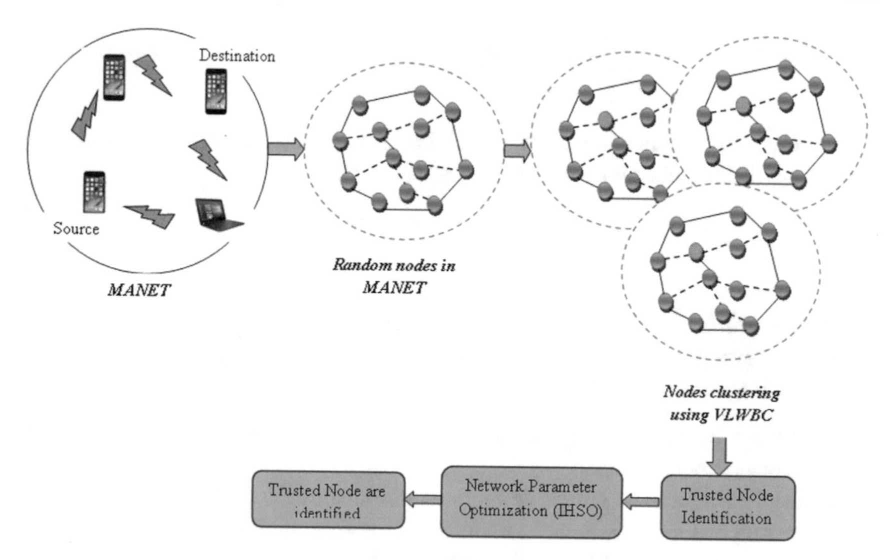

Fig. 1 Structure of proposed work in MANET

Route Handling: This stage handles the produced RREQ packets. Subsequent to setting up the route every one of the packets will be sent to the destination through a similar route. Here, the current RSS estimation of the signal is contrasted with the threshold value. Based on it, transmission intensity of close-by node is diminished in the route reply stage.

Route Termination: It will clarify when the route will end. In the event that, there will be any error in the route then route error packet is sent to the sender and middle nodes will reset transmission intensity of the close-by nodes which thus lessens the aggregate energy utilization by all nodes.

4.2 Cluster Formation Using VLWBC Algorithm

The function of clustering in mobile nodes is to upgrade the network's efficiency, less energy utilization, security, and so forth. The proposed VLWBC decides the weight of each node by utilizing the node's very own features as well as by considering the immediate impact of neighbor node's features [26]. The combined weight of every node is evaluated by the weights of its virtual connections. At long last, among numerous nodes and as indicated by their weights a node is chosen as the cluster head.

The Procedure Involved in VLWBC

Link Neighborhood Contribution: Find the neighbors of each node (s) by defining its degree d_s as

$$d_s = |N(s)| = \sum_{s' \in S, s; \neq s} \{distance(s, s') < node\,transmission\,range\} \quad (1)$$

In order to assign an equal share of neighborhood contribution to each node, 1 is divided by the neighborhood degree, the equation is

$$N_s = \frac{1}{d_s} \quad (2)$$

Link Stability Degree: Due to the mobility of nodes in MANET, nodes may leave their groups and make the groups separate. In this way, a node needs the accompanying capabilities to wind up a cluster head:

- The cluster becomes more stable when the node mobility is slighter than the neighboring node mobility.
- Moving in accordance with neighbors: it diminishes the nodes outward moving and joining different clusters; in addition, if message trade and energy utilization lessens and then the node stability increments;
- The speed of the considered node is the same as the neighboring nodes: a node with these conditions is a possibility to be the cluster head.

Links Consumed Energy: The nodes with the highest residual energy as well as the least consumed energy among their neighbors are selected as the cluster heads. It is calculated by Eq. (3).

$$E_l = \frac{E_{s1} + E_{s2}}{2} \quad (3)$$

The Distance Between Two Linked Nodes: The distance can be calculated by sending packets and measuring the transferring time between the nodes s1 and s2.

$$Dist_{s1,s2} = \sqrt{(X_{s1} - X_{s2})^2 + (Y_{s1} - Y_{s2})^2} \quad (4)$$

Links Combined Weight: The combined weight for a link between nodes are calculated based on four factors such as link neighborhood contribution, link stability, link consumed energy and link distance. The equation is

$$CW_s = w_1 N_s + w_2 Dist_{s1,s2} + w_3 E_l + w_4 link\ stability \quad (5)$$

where, w_1, w_2, w_3, w_4 are the weight factors of each node, X_{s1}, X_{s2} and Y_{s1}, Y_{s2} are the X and Y coordinates of the node s1 and s2 respectively.

4.3 Trusted Node Based Routing

From the clustered nodes, the trusted nodes are selected based on the network parameters such as packet delivering the speed of the node, QoS, network lifetime and energy consumption. A trust model is employed to improve the routing and the performance of MANETs through assessing the trustworthiness of the nodes (communication quality of node) in the networks [27]. The mechanisms involved in trust management are described as below:

The trust module provides secure data transmission in MANET by performing the trust evaluation process of the routing metric, based on the energy factor as in Eq. (6):

$$E_f(s1, s2, t) = mE_f^{SD}(s2, t) + nE_f^{NN}(s2, t) \tag{6}$$

The expansion of Eq. (6) is: $E_f^{SD}(s2, t)$ symbolizes the energy factor of node s1 to s2 by self-detection (SD), $E_f^{NN}(s2, t)$ indicates the energy factor of node s1 to s2 for neighboring nodes (NN). The constant m and n values are in the range of 0–1. By altering the values of m and n, the SD and NN values are changed.

Updating Phase: When a new node enters a MANET, the trust module introduces the energy factor of that node by the condition (6). At the same time, the new node also begins monitoring its neighbors after initializing the energy factor. If the recently joined node is not a malicious node then the trust value will be updated with respect to time. The update phase is responsible for tracking the trust values of the departing and re-entering nodes. An optimal as well as sea cure path from the source to the destination is selected according to the trust feedback information of the intermediate nodes.

4.4 Objective of the Proposed Model

The main objective of our work is to develop a trusted MANET with efficient routing by the network parameter optimization. The fitness function is calculated based on the network parameter by maximizing the QoS, network lifetime and minimizing the energy consumption of nodes. The fitness function is evaluated as the

$$O(v)\ or\ F_j = \{\min(EC), \max(NLT), \max(QoS)\} \tag{7}$$

The desired minimized and maximized value is obtained by the use of inspired optimization technique i.e. Improved HSO algorithm.

Harmony Search (HS) Algorithm: In a standard point of view, the process of optimization can be described to find the best solution of the function from the system within constraints [35, 36, 40]. It is another kind of metaheuristics algorithm impersonating a musicians' approach to deal with discovering harmony while playing music. At the point, when artists attempt to make music, they may utilize one or a mix of the three conceivable techniques for melodic act of spontaneity which are as per the following [29]: (1) playing the first piece, (2) playing in a route like the first piece, and (3) making a piece through arbitrary notes.

Improved HS Algorithm: [28] this model can modify the random value selection i.e. based on the foraging activity of the Krill Herd Optimization (KHO) algorithm in the Eqs. (13, 14) and those steps are explained as below:

Steps Involved in MHSO Algorithm

Initialization: The optimization problem is defined as to achieve the defined objective function $O(v)$, the function is initialized as:

$$f_j \in F_j = 1, 2, \ldots., N \tag{8}$$

where, N denotes the number of decision variables, F_j denotes the set of the possible range of values for each decision variable.

Parameters: Harmony Memory Size (HMS) or the number of solution vectors in the harmony memory; Harmony Memory considering Rate (HMCR), HMCR $\in$ [0, 1], Pitch Adjusting Rate (PAR) $\in$ [0, 1].

Harmony Memory Initialization and Evaluation: A random initial population is generated, such as:

$$f_{i,j}^{0} = f_j^{\min} + r_j(f_j^{\max} - f_j^{\min}) \tag{9}$$

where, i = 1, 2 ....HMS, j = 1, 2 ....N and $r_j \in [0, 1]$ is a uniformly distributed random number generated new for each value of j. Solution vectors in HM are analyzed, and their objective function values are then calculated.

Improvise a New Harmony: A New Harmony vector is produced in view of three guidelines, for example, memory consideration, pitch adjustment, and random selection. In the memory consideration, the value of the first decision variable N_1^1 for the new vector is selected over any of the values in the specified HM ($Z_1^1 - Z_1^{HMS}$) range. HM is similar to the step where the musician utilizes memory to produce a tune. Values of the other decision variables are selected in the same manner. The HMCR $\in$ [0, 1] is the rate of choosing one value from the historical one stored in the HM, while $(1 - HMCR)$ is the rate of randomly selecting one value from the desired range of values.

$$Z_i' = \begin{cases} Z_i' \in \{Z_i^1, Z_i^2 \ldots . Z_i^{HMS}\ with\ probability\ HMCR \\ Z_i' \in Z\ with\ probability\ (1 - HMCR) \end{cases} \tag{10}$$

For instance, an HMCR of 0.90 shows that the HS will pick the decision variable incentive from historically stored values in HM with a 90% probability or from the whole conceivable range with a (100–90) % probability. Each part acquired by the memory consideration is inspected to decide if it ought to be pitch-adjusted. This activity utilizes the PAR $\in$ [0, 1] parameter, which is the rate of pitch adjustment as takes after:

$$Z_i' = \begin{cases} Adjusting\ pitch\ with\ probability \\ Doing\ nothing\ with\ probability\,(1 - PAR) \end{cases} \tag{11}$$

The value of (1-PAR) sets the rate of doing nothing. If the pitch adjustment decision for N_i' is YES, N_i' is replaced as follows:

$$Z_i' = Z_i' \pm rand \times bw \tag{12}$$

where, bw is an arbitrary distance bandwidth, $rand$ is a random number and its selection is based on the foraging activity of KHO algorithm.

Foraging Motion of KHO: The foraging motion is formulated in terms of two main effective parameters. The first one is the food location and the second one is the previous experience of the food location. This motion can be expressed for the ith krill individual as follows:

$$B_i = V_m \delta_i + \omega_m B_i^{old} \tag{13}$$

where

$$\delta_i = \delta_i^{food} + \delta_i^{best} \tag{14}$$

Here the scavenging velocity is denoted by V_m, the dormancy weight of the searching movement is ω_m [0, 1], $B_i \rightarrow$ Foraging motion, the best food searching and best fitness of the krill are δ_i^{food} and δ_i^{best} respectively.

Update Harmony Memory: For each new value of harmony the value of the objective function, calculated. If the New Harmony vector is superior to the worst harmony in the HM, the New Harmony is included in the HM and the existing worst harmony is excluded from the HM. If the maximum number of iteration is reached, then terminate the IHSO process. The graphical representation of IHSO is shown in Fig. 2.

4.5 Trusted Node

Based on the attained optimal parameters i.e. minimized Energy consumption, maximized QoS and network lifetime, the trusted node with its routing path is selected. In route discovery process, the trust module helps the destination node to initialize the threshold value for the energy factor and it also examines the energy factor of the nodes to include or reject the nodes in the route for communication.

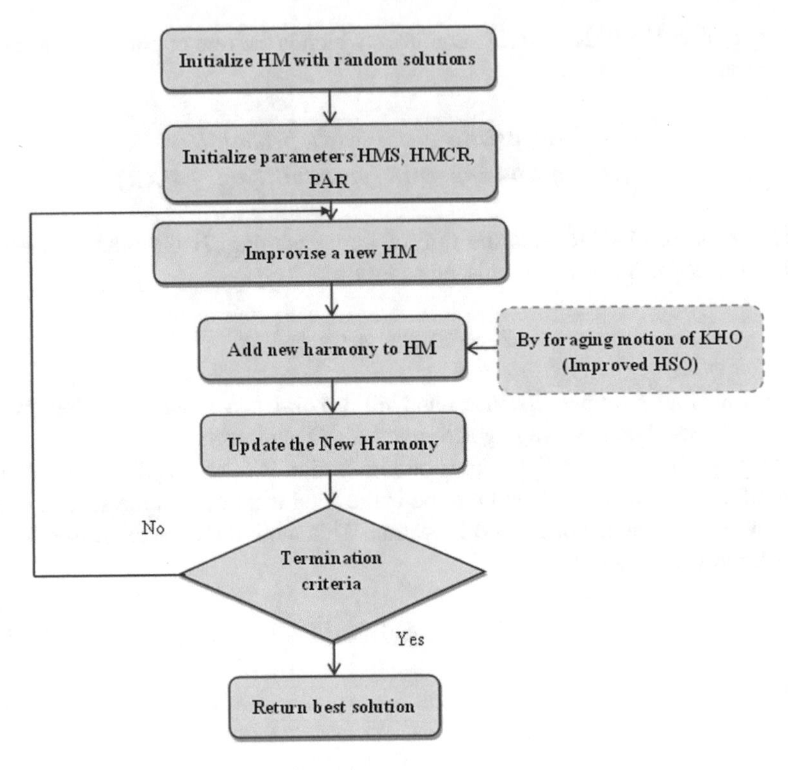

Fig. 2 Graphical representation of IHSO

5 Result Analysis

In the result analysis part, the performance of the proposed energy efficient trusted node identification of MANET model is investigated by examining the network parameters in terms of energy consumption, packet delivery ratio, network lifetime and QoS. The execution of the proposed study is analyzed with the help of Network Simulation Tool-Version 2 (NS2). The simulation results of the proposed work are described in this section.

5.1 Performance Measures

Energy Consumption (EC): Energy utilization is very much characterized as the communication overhead of the nodes where a firm number of false information are infused into a network.

Table 1 Results for proposed VLWBC-IHSO model

Number of mobile nodes	PDR (%)	NLT (h)	EC (J)	QoS	Clustering efficiency
100	97	266	114	95	96
200	88	139	127	94.9	94
300	82	141	204	94.7	92.86
400	84	111	123	94.5	91.99
500	80	99	144	94.4	93.55

Packet to Delivery Ration (PDR): The amount of information which was viably transferred to the end node related to the measure of information which was transferred by the transmitter.

Network Life Time (NLT): Network lifetime is defined as the time until the first node's energy runs out is an important performance metric in WSNs. In this work, the connection time between mobile nodes has been controlled to maximize the network lifetime.

Quality of Service: The overall performance of the mobile network services are defined as QoS., for example, the performance of a service such as a telephony or a computer network or a cloud computing service, in specific the performance observed by the users of the mobile network.

Table 1 explains the result analysis of a proposed model for a different number of mobile nodes. Based on the node quality, mobile nodes are clustered using VLWBC model. The performance of the proposed model is analyzed in terms of PDR, NLT, EC, QoS, and clustering efficiency. Compared to existing models, the maximum clustering efficiency is attained in the VLWBC-IHSO model i.e. 94.25% for 100 numbers of nodes. Similarly, the performance measures are analyzed for 200, 300, 400 and 500 number of mobile nodes.

Performance Analysis of Clustered Nodes

Figure 3a demonstrates the Packet to delivery Ratio (PDR) and Fig. 3b explains the energy consumption for various nodes. These parameters are analyzed and compared with three different algorithms such as BPSO, KHO, and the proposed IHSO. On comparing the proposed model with the existing approach, IHSO attains high PDR i.e. 20 at the node 100. Figure 3b explains the energy consumption of each mobile node during the data transmission from the source node to the destination. The proposed IHSO algorithm needs minimum energy for data transmission compared to other existing algorithms.

Figure 3c describes the Network lifetime (h) for different nodes based optimization techniques. At node 100, the KHO algorithm achieves the network life time as 125.63 h, constrained technique as 90.3 h; the proposed IHSO algorithm reaches NLT as 270 h. Similarly, for other nodes (200–500) the NLT is analyzed and compared. Finally, the graph concludes the proposed approach attains maximum lifetime in the MANETs. Figure 3d shows the quality of service which clearly depicts the node

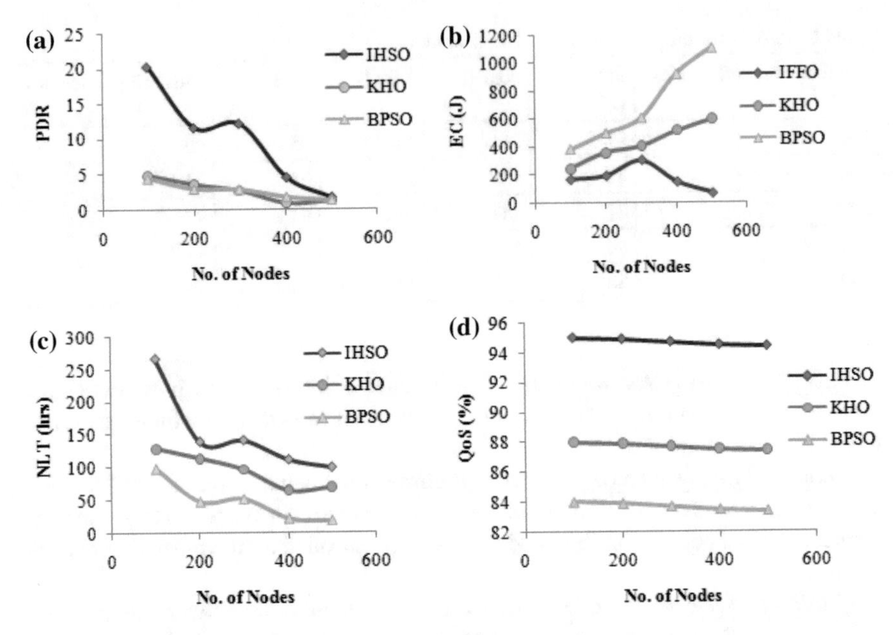

Fig. 3 Performance analysis of nodes using algorithms

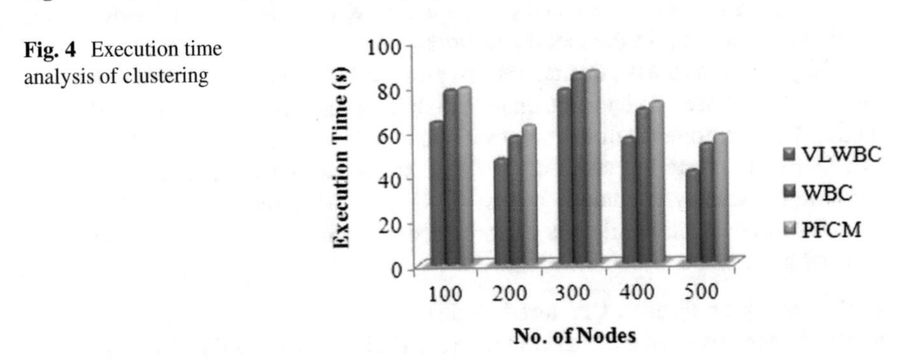

Fig. 4 Execution time analysis of clustering

quality by optimizing the network parameters using IHSO algorithm. Compared to existing optimization, IHSO attains high QoS in the range of 94–95%.

Execution Time for Clustering Algorithms

Figure 4 illustrates the time taken for clustering the mobile nodes in MANETs with the use of clustering techniques like VLWBC, WBC, and PFCM. For node 100, the VLWBC algorithm takes 64 s to cluster the random nodes in the network, WBC takes 79 s and PFCM algorithm takes 78 s to cluster the nodes. Similarly, the clustering time is analyzed for the other nodes such as 200, 300, 400 and 500 and it is shown in the bar graph. The proposed VLWBC algorithm attains minimum time to cluster the nodes compared to WBC and PFCM approach.

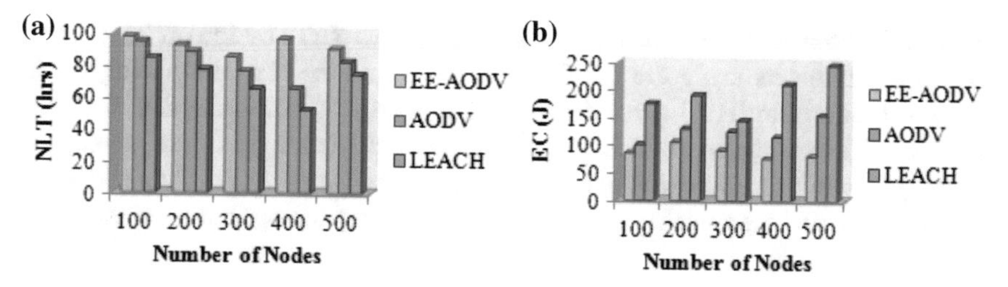

Fig. 5 Performance analysis of routing protocol **a** NLT and **b** EC

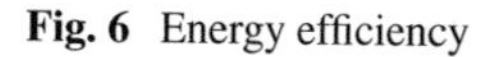
Fig. 6 Energy efficiency

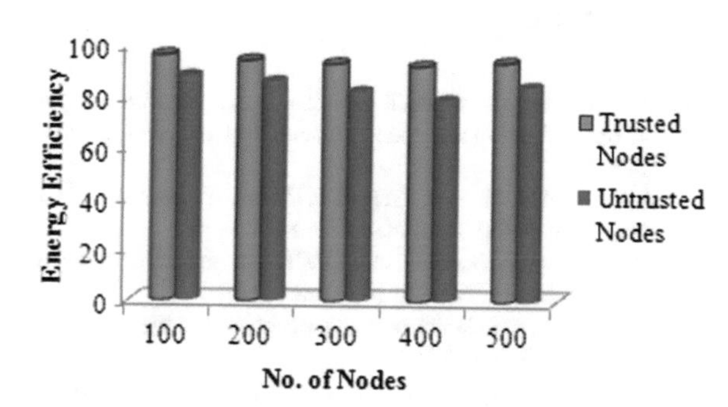

Performance Analysis of Routing Protocol
The performance analysis of different routing protocol is explained in Fig. 5. The lifetime and the energy consumption of nodes are illustrated in the figure. The analyzed protocols are EE-AODV, AODV, and LEACH. Compared to LEACH and AODV, the proposed EE-AODV protocol achieves maximum network lifetime and minimum energy consumption for all mobile nodes in MANETs.

Energy Efficiency of Trusted Nodes
The Energy Efficiency analysis of the both trusted and un-trusted mobile network is analyzed and the efficiency is compared in Fig. 6. During the data transmission, trusted nodes require less time and also it consumes low energy to transfer the data from one node to another. By identifying the trusted nodes in MANETs an efficient mobile to a mobile communication is accomplished.

6 Conclusion

This paper presented the trusted node identification of MANETs with maximizing the energy efficiency, network lifetime and QoS. These objectives were attained by the link based clustering algorithm i.e. VLWBC algorithm. It clusters the mobile nodes in different rounds and selecting any nodes as cluster heads in some rounds, it has been able to reduce the number of transmitted messages from each node to different nodes and to the base station, saving more energy in the network. Then the trusted mobile node was obtained among the clustered nodes in MANETs by the optimization of network parameters using the IHSO algorithm. From the simulation analysis, the proposed algorithm IHSO gives the minimum energy consumption and maximum NLT and QoS, compared to the KHO and BPSO technique. In the future, we can also extend this work by improving the energy efficiency in terms of minimized cost and maximized QoS by proposing more efficient clustering and optimization methods.

References

1. W.A. Jabbar, M. Ismail, R. Nordin, Energy and mobility conscious multipath routing scheme for route stability and load balancing in MANETs. Simul. Model. Pract. Theory **77**, 245–271 (2017)
2. A. M. Shabut, M. S. Kaiser, K. P. Dahal, W. Chen (2018). A multidimensional trust evaluation model for MANETs. J. Netw. Comput. Appl.
3. M. Sakthivel, V.G. Palanisamy, Enhancement of accuracy metrics for energy levels in MANETs. Comput. Electr. Eng. **48**, 100–108 (2015)
4. S. Suganya, S. Palaniammal, An optimized energy consumption algorithm for MANETs. Proc. Eng. **38**, 903–910 (2012)
5. S. Sarkar, R. Datta, A trust based protocol for energy-efficient routing in self-organized MANETs, in 2012 Annual IEEE India Conference (INDICON) (IEEE), pp. 1084–1089
6. C.N. Kumar, N. Satyanarayana, Multipath QoS routing for traffic splitting in MANETs. Proc. Comput. Sci. **48**, 414–426 (2015)
7. R.K. Bar, J.K. Mandal, M.M. Singh, QoS of MANet through trust based AODV routing protocol by exclusion of black hole attack. Proc. Technol. **10**, 530–537 (2013)
8. K. Badhe, S. Jain, Implementation of mobility and QoS aware energy efficient anycast routing in MANET. Int. J. Advanc. Netw. Appl. **9**(1), 3338 (2017)
9. H.S. Jassim, S. Yussof, T.S. Kiong, S.P. Koh, R. Ismail, A routing protocol based on trusted and shortest path selection for mobile ad hoc network, in 2009 IEEE 9th Malaysia International Conference on 2009, Communications (MICC), December (IEEE), pp. 547–554
10. R.S. Mangrulkar, P.V. Chavan, S.N. Dagadkar, Improving route selection mechanism using trust factor in AODV routing protocol for MANeT. Int. J. Comput. Appl. 0975–8887, 36–39
11. S. Biswas, Dey, P., Neogy, S., Secure checkpointing-recovery using trusted nodes in MANET, in 2013 4th International Conference on Computer and Communication Technology (ICCCT), 2013 September (IEEE) pp. 174–180
12. V. Jayalakshmi, A.T. Razak, Selection of trusted nodes in MANET by using analytic network process. Compusoft **3**(3), 657 (2014)
13. J.H. Cho, A. Swami, R. Chen, Modeling and analysis of trust management with trust chain optimization in mobile ad hoc networks. J. Netw. Comput. Appl. **35**(3), 1001–1012 (2012)
14. J. Sinha, C. Choudhary, C., Identification of trusted nodes in mobile Adhoc network. J. Sci. Res. **3**(2), 2137–2141
15. R.R. Ravi, V. Jayanthi, Energy efficient neighbor coverage protocol for reducing rebroadcast in MANET. Proc. Comput. Sci. **47**, 417–423 (2015)
16. S. Divya, B. Murugesakumar, Energy efficient routing protocols in mobile Ad Hoc network (MANET)–a review. Int. J. Sci. Res. (IJSR) **2**(1), 292–297 (2013)
17. B. Narasimhan, R. Balakrishnan, Energy Efficient Ad-Hoc On-Demand Distance Vector (Ee-Aodv) routing protocol for mobile ad hoc networks. Int. J. Sci. Res. Comput. Sci. **4**(9), 66–69 (2013)
18. S.P. Jaladi, Ant colony optimization based routing to improve QoS in MANETs. J. Comput. Sci. Informat. Technol. **6**(1), 609–613 (2015)

19. S. Chavhan, P. Venkataram, Emergent intelligence based QoS routing in MANET. Proc. Comput. Sci. **52**, 659–664 (2015)
20. S. Venkatasubramanian, N. Gopalan, Improving energy efficient QOS performance for heterogeneous MANET. Int. J. Comput. Netw. Commun. Secur. **2**(2), 70–81
21. W.K. Kuo, S.H. Chu, Energy efficiency optimization for mobile ad hoc networks. IEEE Access **4**, 928–940 (2016)
22. S. Sridhar, R. Baskaran, P. Chandrasekar, Energy supported AODV (EN-AODV) for QoS routing in MANET. Proc. Soc. Behavior. Sci. **273**, 294–301 (2013)
23. V.V. Mandhare, V.R. Thool, R.R. Manthalkar, QoS Routing enhancement using metaheuristic approach in mobile ad-hoc network. Comput. Netw. **110**, 180–191 (2016)
24. S. Jamali, L. Rezaei, S.J. Gudakahriz, An energy-efficient routing protocol for MANETs: a particle swarm optimization approach. J. Appl. Res. Technol. **11**(6), 803–812 (2013)
25. H. Ashwini, A. Rajashri, Implementation of energy efficient AODV protocol for MANET. J. Eng. Res. Technol. **4**(05), 1577–1580
26. A. Karimi, A. Afsharfarnia, F. Zarafshan, S.A.R. Al-Haddad, A novel clustering algorithm for mobile ad hoc networks based on determination of virtual links' weight to increase network stability. Scientif. World J. (2014)
27. M.S. Pathan, N. Zhu, J. He, Z.A. Zardari, M.Q. Memon, M.I. Hussain, An efficient trust-based scheme for secure and quality of service routing in MANETs. Future Internet **10**(2), 16 (2018)
28. M. Mahdavi, M. Fesanghary, E. Damangir, An improved harmony search algorithm for solving optimization problems. Appl. Math. Comput. **188**(2), 1567–1579 (2007)
29. J.H. Kim, Harmony search algorithm: a unique music-inspired algorithm. Proc. Eng. **154**, 1401–1405 (2016)
30. M. Elhoseny, A.E. Hassanien, Secure data transmission in WSN: an overview, in *Dynamic Wireless Sensor Networks. Studies in Systems, Decision and Control*, vol. 165. Springer, Cham, pp. 115–143 (2019). https://doi.org/10.1007/978-3-319-92807-4_6
31. M. Elhoseny, H. Elminir, A. Riad, X. Yuan, A secure data routing schema for WSN using elliptic curve cryptography and homomorphic encryption. J. King Saud Univ. Comput. Informat. Sci. Elsevier **28**(3):262–275 (2016). (http://dx.doi.org/10.1016/j.jksuci.2015.11.001)
32. X. Yuan, M. Elhoseny, H. K El-Minir, A.M. Riad, A genetic algorithm-based, dynamic clustering method towards improved WSN longevity. J. Netw. Syst. Manag. Springer US, **25**(1):21–46 (2017). https://doi.org/10.1007/s10922-016-9379-7
33. M. Elhoseny, A. Farouk, N. Zhou, M.-M. Wang, S. Abdalla, J. Batle, Dynamic multi-hop clustering in a wireless sensor network: performance improvement. Wireless Personal Commun. Springer US, **95**(4), 3733–3753. https://doi.org/10.1007/s11277-017-4023-8
34. M. Elhoseny, A. Tharwat, A. Farouk, A.E. Hassanien, K-Coverage model based on genetic algorithm to extend WSN lifetime. IEEE Sens. Lett. **1**(4), 1–4. IEEE (2017). (https://doi.org/10.1109/lsens.2017.2724846)
35. K. Shankar, P. Eswaran, RGB based multiple share creation in visual cryptography with aid of elliptic curve cryptography. China Communications **14**(2), 118–130 (2017)
36. K. Shankar, P. Eswaran, RGB-based secure share creation in visual cryptography using optimal elliptic curve cryptography technique. J. Circ. Syst. Comput. **25**(11), 1650138 (2016)
37. M. Elhoseny, A.E. Hassanien, Extending homogeneous WSN lifetime in dynamic environments using the clustering model, in *Dynamic Wireless Sensor Networks. Studies in Systems, Decision and Control*, vol. 165 (Springer, Cham, 2019), pp. 73–92. https://doi.org/10.1007/978-3-319-92807-4_4)
38. M. Elhoseny, A.E. Hassanien, Optimizing cluster head selection in WSN to prolong its existence, in *Dynamic Wireless Sensor Networks. Studies in Systems, Decision and Control*, vol. 165 (Springer, Cham, 2019), pp. 93–111. https://doi.org/10.1007/978-3-319-92807-4_5
39. W. Elsayed, M. Elhoseny, S. Sabbeh, A. Riad, Self-maintenance model for wireless sensor networks. Comput. Electr. Eng. In Press, Available Online Dec 2017. https://doi.org/10.1016/j.compeleceng.2017.12.022)
40. S.K. Lakshmanaprabu, K. Shankar, A. Khanna, D. Gupta, J.J. Rodrigues, P.R. Pinheiro, V.H.C. De Albuquerque, Effective features to classify big data using social internet of things. IEEE Access **6**, 24196–24204 (2018)

Printed in the United States
By Bookmasters